GRAPHING CALCULATOR MANUAL

TRIGONOMETRY GRAPHS AND MODELS

MARVIN L. BITTINGER
Indiana University--Purdue University at Indianapolis

JUDITH A. BEECHER
Indiana University--Purdue University at Indianapolis

DAVID ELLENBOGEN
St. Michael's College

JUDITH A. PENNA
Indiana University--Purdue University at Indianapolis

JUDITH A. PENNA
Indiana University--Purdue University at Indianapolis

ADDISON-WESLEY

An imprint of Addison Wesley Longman, Inc.

Reading, Massachusetts • Menlo Park, California • New York • Harlow, England
Don Mills, Ontario • Sydney • Mexico City • Madrid • Amsterdam

Reproduced by Addison Wesley Longman from camera-ready copy supplied by the author.

ISBN 0-201-34568-4

1 2 3 4 5 6 7 8 9 10 VG 00 99 98 97

Table of Contents

The TI-82 and TI-83 Graphics Calculators

Preliminaries

Press ON to turn on the TI-82 or TI-83 graphing calculator. (ON is the key at the bottom left-hand corner of the keypad.) The display contrast can be adjusted by first pressing 2nd. (2nd is the blue key on the TI-82 and the yellow key on the TI-83 in the left column of the keypad.) Then press and hold △ to increase the contrast or ▽ to decrease the contrast. To turn the grapher off, press 2nd OFF. (OFF is the second operation associated with the ON key.) The grapher will turn itself off automatically after about five minutes without any activity.

It will be helpful to read the Getting Started section of your grapher Guidebook before proceeding. See pages 1 - 14 of the TI-82 Guidebook or pages 1 - 18 of the TI-83 Guidebook.

Press MODE to display the MODE settings. Initially you should select the settings on the left side of the display.

To change a setting use ▽ or △ to move the cursor to the line of that setting. Then use ▷ or ◁ to move the blinking cursor to the desired setting and press ENTER. Press CLEAR or 2nd QUIT to leave the MODE screen. (QUIT is the second operation associated with the MODE key.)

Chapter G
Introduction to Graphs and Graphers

SETTING THE VIEWING WINDOW

The viewing window is the portion of the coordinate plane that appears on the grapher's screen. It is defined by the minimum and maximum values of x and y: Xmin, Xmax, Ymin, and Ymax. The notation [Xmin, Xmax, Ymin, Ymax] is used in the text to represent these window settings or dimensions. For example, [−12, 12, −8, 8] denotes a window that displays the portion of the x-axis from −12 to 12 and the portion of the y-axis from −8 to 8. In addition, the distance between tick marks on the axes is defined by the settings Xscl and Yscl. The TI-83 includes an additional setting, Xres, which sets the pixel resolution. We usually select Xres = 1. The window corresponding to the settings [−20, 30, −12, 20], Xscl = 5, Yscl = 2, (Xres = 1), is shown below.

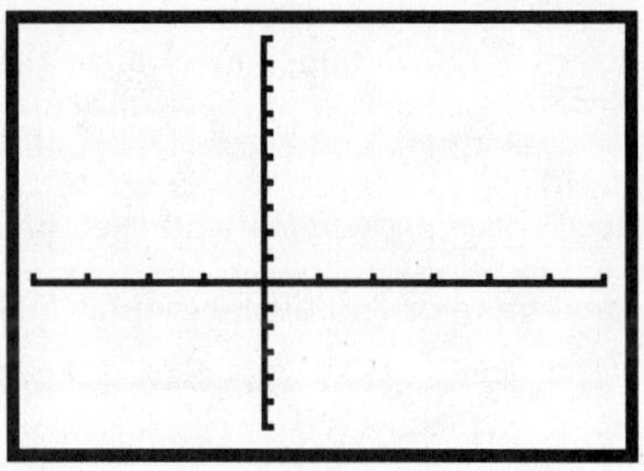

Press the WINDOW key on the top row of the keypad to display the current window settings on your grapher. The standard settings are shown below.

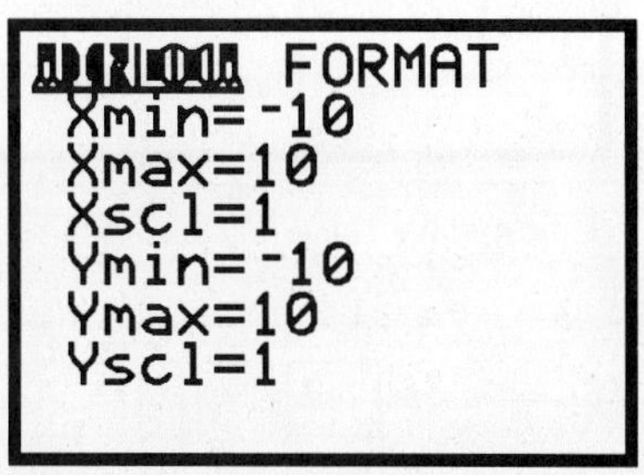

To change a setting press ▽ to move the cursor to the setting you wish to change and enter the new value. For example, to change from the standard settings to [−20, 30, −12, 20], Xscl = 5, Yscl = 2, on the TI-82 press ▽ (−) 2 0 ENTER 3 0 ENTER 5 ENTER (−) 1 2 ENTER 2 0 ENTER 2 ENTER. You must use the gray (−) key on the bottom row of the keypad rather than the dark blue − key in the right-hand column to enter a negative number. (−) represents "the opposite of" or "the additive inverse of" whereas − is the subtraction key. The ▽ key may be used instead of ENTER after typing each window setting. On the TI-83, the cursor appears at Xmin after WINDOW is pressed, so the ▽ preceding −20 above should be omitted.

To return quickly to the standard window setting [−10, 10, −10, 10], Xscl = 1, Yscl = 1, press [ZOOM] 6.

GRAPHING EQUATIONS

An equation must be solved for y before it can be graphed on the TI-82 and the TI-83.

Example 8 (a), page 8 (Page numbers refer to pages in the textbook.): To graph $2x+3y=18$, first solve for y, obtaining $y=\dfrac{18-2x}{3}$. Then press [Y =], the key at the top left-hand corner of the keypad. If there is currently an expression displayed for Y_1, press [CLEAR] to delete it. Do the same for expressions that appear on all other lines by using [▽] to move to a line and then pressing [CLEAR]. Then use [△] to move the cursor to the top line beside "Y_1 =." Now press [(] 18 [−] 2 [X, T, Θ] [)] [÷] 3 to enter the right-hand side of the equation. (The key that produces the X on the TI-83 is marked [X, T, Θ, n].) Note that without the parentheses the expression $18-\dfrac{2x}{3}$ would have been entered.

You can edit your entry if necessary. If, for instance, you pressed 5 instead of 8, use the [◁] key to move the cursor to 5 and then press 8 to overwrite it. If you forgot to type the right parenthesis, move the cursor to the division symbol /; then press [2nd] [INS] [)] to insert the parenthesis before the division symbol. ([INS] is the second operation associated with the [DEL] key.) You can continue to insert symbols immediately after the first insertion without pressing [2nd] [INS] again. If you typed 25 instead of 2, move the cursor to 5 and press [DEL]. This will delete the 5.

Once the equation is entered correctly, select a viewing window and then press [GRAPH] to display the graph. You may change the viewing window as desired to reveal more or less of the graph. The standard window is shown here.

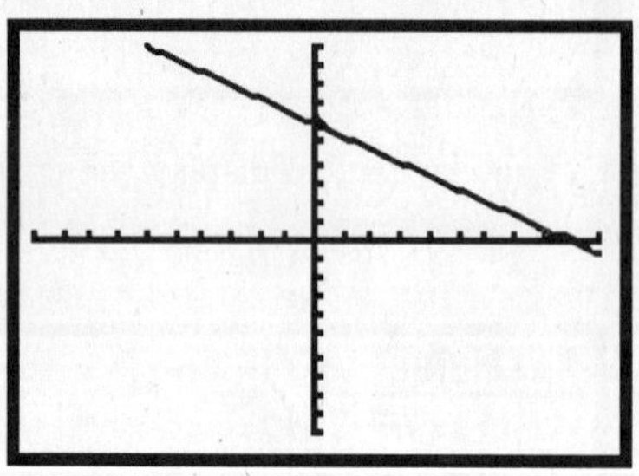

Example 8(c), page 8: To graph $x=y^2+1$, first solve the equation for y : $y=\pm\sqrt{x-1}$. To obtain the entire graph of $x=y^2+1$, you must graph $y_1=\sqrt{x-1}$ and $y_2=-\sqrt{x-1}$ on the same screen. Press [Y =] and clear any expressions that currently appear. With the cursor beside "Y_1 =" press [2nd] [√] [(] [X, T, Θ] [−] 1 [)]. ([√] is the second operation associated with the [x^2] key.) Note that if the parentheses had not been used, the equation entered would have been $y_1=\sqrt{x}-1$. On the TI-83, the left parenthesis appears along with the radical symbol, so a separate keystroke is not necessary to introduce it.

Now use [▽] to move the cursor beside "Y2 =." There are two ways to enter $y_2=-\sqrt{x-1}$. One is to enter the expression $-\sqrt{x-1}$ directly by pressing [(−)] [2nd] [√] [(] [X, T, Θ] [−] 1 [)].

The other method of entering y_2 is based on the observation that $-\sqrt{x-1}$ is the opposite of the expression for y_1. That is, $y_2 = -y_1$. To enter this on the TI-82, place the cursor beside "Y2 =" and press [(−)] [2nd] [Y-VARS] 1 1. (Y-VARS is the second operation associated with the [VARS] key.) This enters the opposite of y_1 as the expression for y_2. On the TI-83, press [(−)] [VARS] [▷] to select Y-Vars. Then press 1 1 to select y_1.

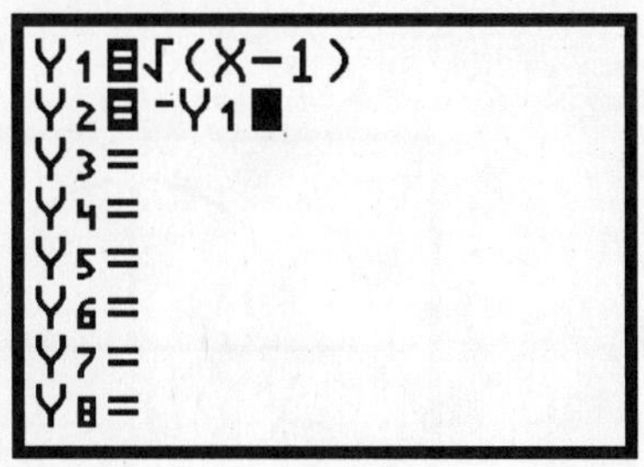

Select a viewing window and press [GRAPH] to display the graph. The window shown here is $[-2, 5, -5, 5]$, Xscl = 1, Yscl = 1.

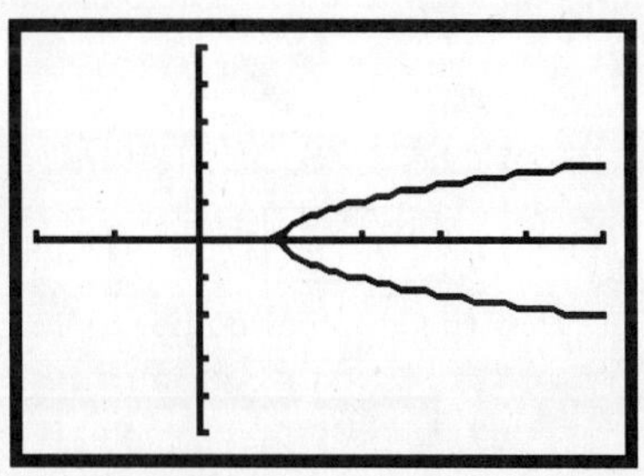

The top half is the graph of y_1, the bottom half is the graph of y_2, and together they yield the graph of $x = y^2 + 1$.

SQUARING THE VIEWING WINDOW

In the standard window, the distance between tick marks on the y-axis is about 2/3 the distance between tick marks on the x-axis. It is often desirable to choose window dimensions for which these distances are the same, creating a "square" window. Any window in which the ratio of the length of the y-axis to the length of the x-axis is 2/3 will produce this effect.

This can be accomplished by selecting dimensions for which Ymax − Ymin $= \frac{2}{3}$(Xmax − Xmin). For example, the windows $[-12, 12, -8, 8]$ and $[-6, 6, -4, 4]$ are square. To illustrate this, graph the circle $x^2 + y^2 = 49$ in the standard window by first entering $y_1 = \sqrt{49 - x^2}$ and $y_2 = -\sqrt{49 - x^2}$ or $y_2 = -y_1$. Note that x^2 can be entered either by pressing

X, T, Θ x^2 or by pressing X, T, Θ ∧ 2. The ∧ key can be used to enter any exponent, but for the exponent 2 the x^2 key is more efficient. Select ZOOM 6 if the window settings are not already standard or press GRAPH if the standard settings have previously been entered.

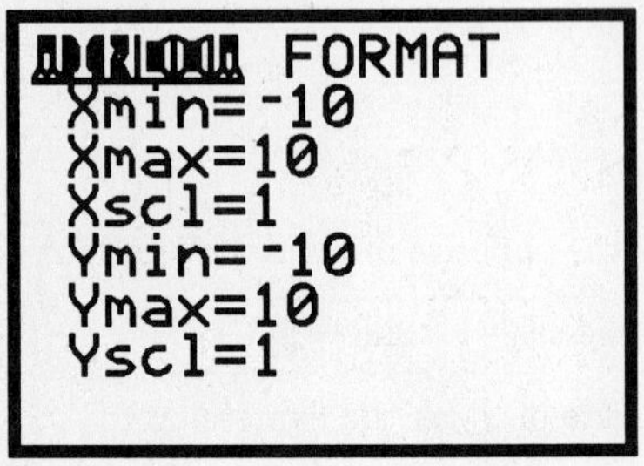

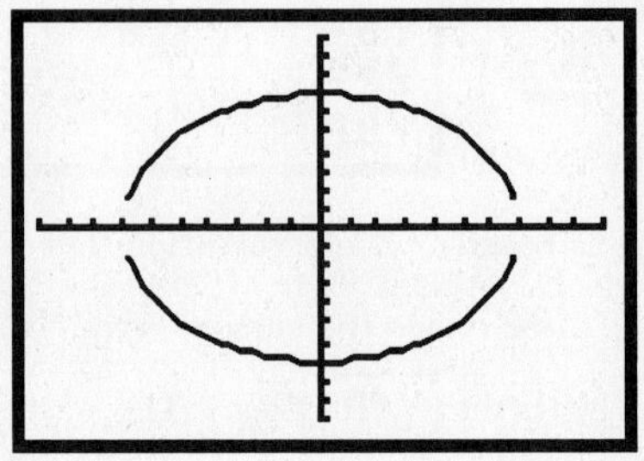

Note that the graph does not appear to be a circle.

Now change the window dimensions to $[-12, 12, -8, 8]$, Xscl = 1, Yscl = 1, and press GRAPH.

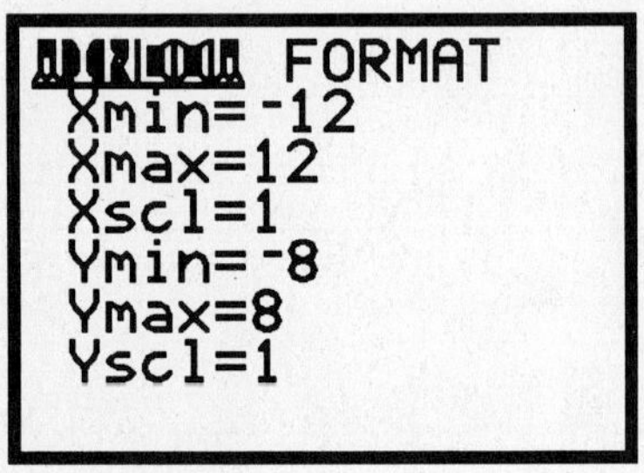

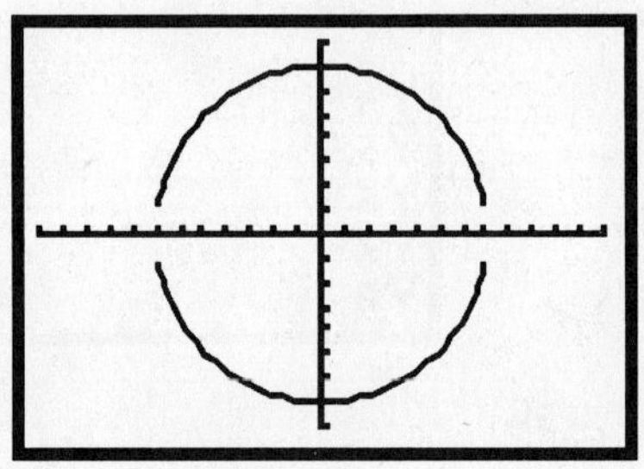

Observe that the distance between tick marks appears to be the same on both axes and that the graph appears to be a circle.

The window can also be squared using the grapher's ZSquare feature. Press ZOOM 6 to return to the graph of $x^2 + y^2 = 49$ in the standard window. Now press ZOOM 5 to select the ZSquare feature. The resulting window dimensions and graph are shown below. Note that the graph also appears to be a circle in this window.

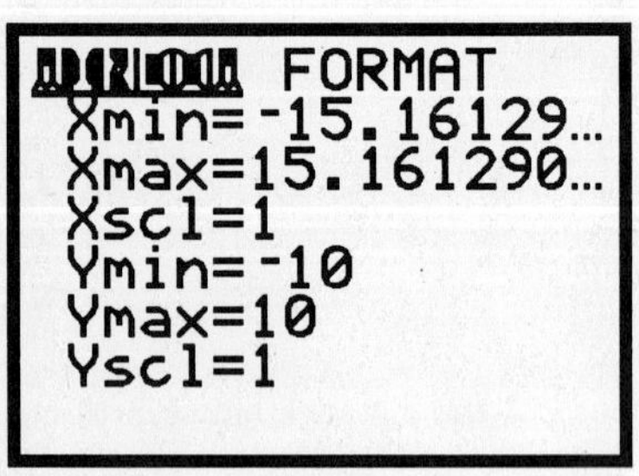

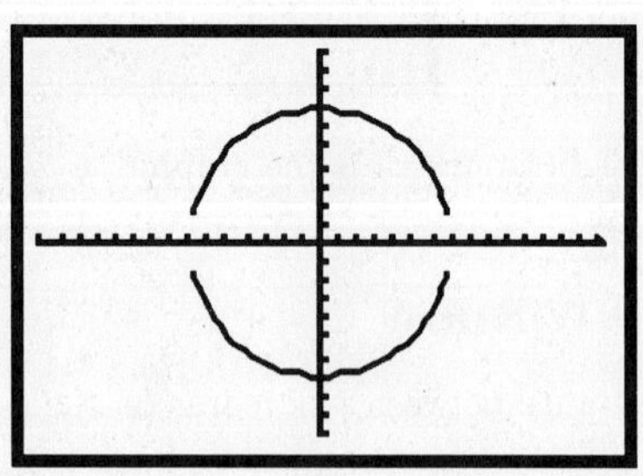

THE TABLE FEATURE

For an equation entered in the "Y =" screen, a table of x-and y-values can be displayed. For example, on the "Y =" screen enter $y_1 = 3x^3 - 5x^2 + 2x - 1$ by positioning the cursor beside "Y_1 =" and pressing 3 X, T, Θ ∧ 3 − 5 X, T, Θ x^2 + 2 X, T, Θ − 1. Then press 2nd TblSet to display the table set-up screen. (TblSet is the second function associated with the WINDOW key.) A minimum value of x can be chosen along with an increment for the x-values. Press -5 ▽ .1 to select a minimum x-value of -5 and an increment of 0.1. The "Indpnt" and "Depend" settings should both be "Auto." If either is not, use the ▽ key to position the blinking cursor over "Auto" on that line and then press ENTER. To display the table press 2nd TABLE. (TABLE is the second function associated with the GRAPH key.)

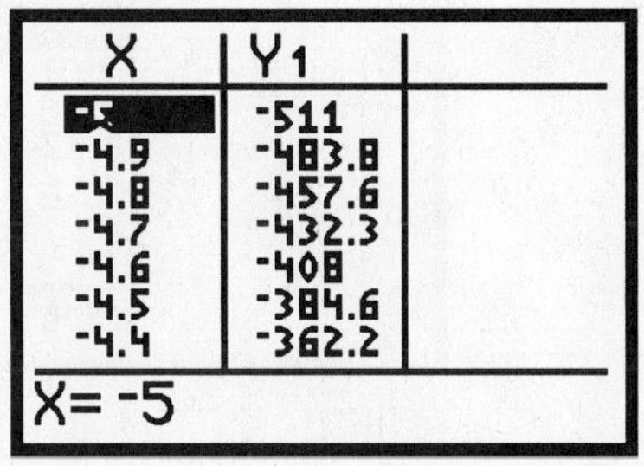

Use the ▽ and △ keys to scroll through the table. For example, by using ▽ to scroll down we can see that $y_1 = -213$ when $x = -3.6$. Using △ to scroll up, observe that $y_1 = -2530$ when $x = -8.9$.

THE SPLIT SCREEN

A horizontally split screen can be used on both the TI-82 and the TI-83 to display the graph of an equation along with its corresponding table of values. The TI-83 will also display a graph and a table on a vertically split screen. To produce a split screen, we first use the MODE menu. For instance, for the table settings and the equation $y = 3x^3 - 5x^2 + 2x - 1$ entered as above, select a viewing window. Then on the TI-82 press MODE ▽ ▽ ▽ ▽ ▽ ▽ ▷ ENTER. This selects the split screen option.

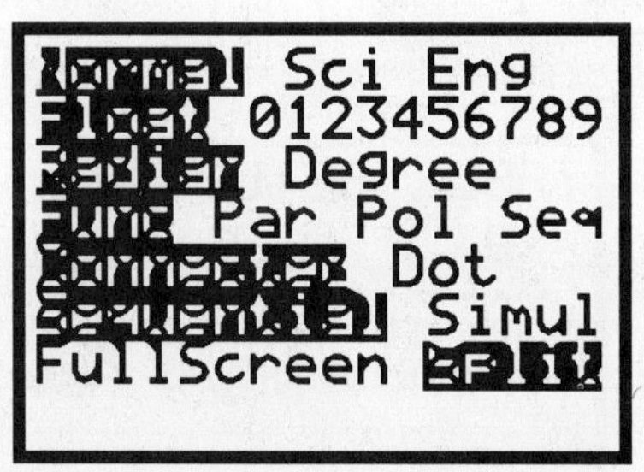

On the TI-83 press MODE ▽ ▽ ▽ ▽ ▽ ▽ ▽ ▷ ENTER to select a horizontally split screen or press MODE ▽ ▽ ▽ ▽ ▽ ▽ ▷ ▷ ENTER to select a vertically split screen.

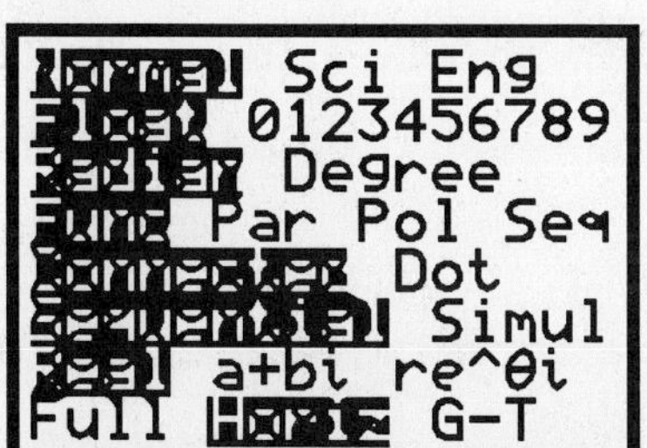

Now press 2nd TABLE. In the horizontal mode the result is a split screen displaying the graph at the top with two rows of the table below it. In the vertical mode the graph is displayed on the left with seven rows of the table to its right. The ▽ and △ keys can be used to scroll through the table as before.

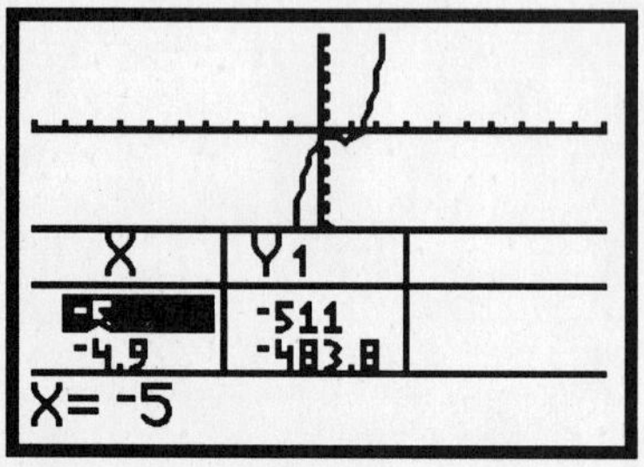

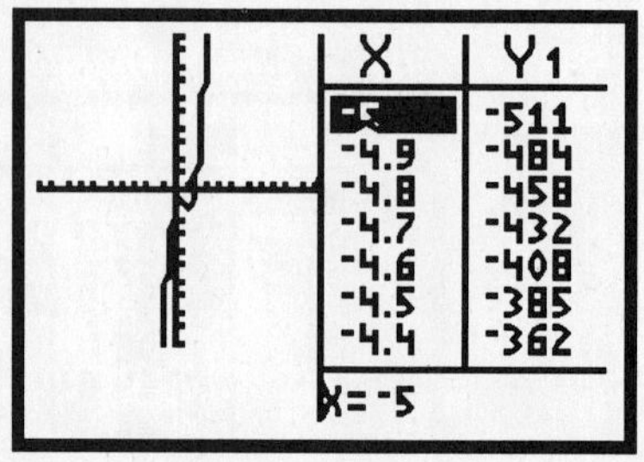

Return the TI-82 to full screen mode by pressing MODE ▽ ▽ ▽ ▽ ▽ ▽ ENTER. On the TI-83 press MODE ▽ ▽ ▽ ▽ ▽ ▽ ▽ ENTER.

The TABLE feature can also be used to evaluate an expression. Enter $y_1 = 5x^4 - 6x^2 + 4$ in the "Y =" screen. Then press 2nd TblSet ▽ ▽ ▷ ENTER to set the table in ASK mode. In ASK mode the grapher disregards the values of TblMin and Δ Tbl.

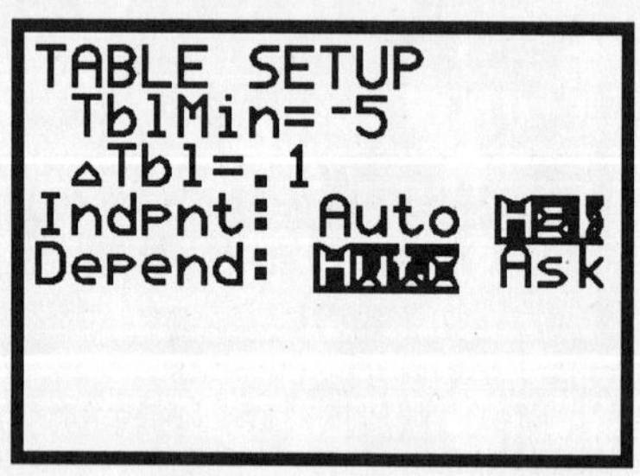

Press 2nd TABLE and an empty table is displayed. Now x-values can be entered in the X-column and the corresponding y-values will be displayed in the Y_1-column. For example, when (−) 9 ENTER is pressed, −9 appears in the X-column and the grapher computes and enters 32323 in the Y_1-column. This is the value of $5x^4 - 6x^2 + 4$ when $x = -9$, or $5(-9)^4 - 6(-9)^2 + 4$. Press 16 ENTER and 326148 appears in the Y_1-column. This is the value of the expression when $x = 16$. You can continue to enter x-values as desired.

IDENTITIES

An equation that is true for every possible real-number substitution for the variable is an identity. The grapher can be used to provide a partial check whether an equation is an identity. Either a graph or a table can be used to do this.

Example 9 (a), page 11: Determine whether $(x^2)^3 = x^6$ appears to be an identity.

To determine whether this equation appears to be an identity, graph $y_1 = (x^2)^3$ and $y_2 = x^6$. Examine the graphs in several viewing windows. The graphs appear to coincide no matter what the window. Thus, although there is a possibility that the graphs fail to coincide outside the windows that were examined, the equation appears to be an identity.

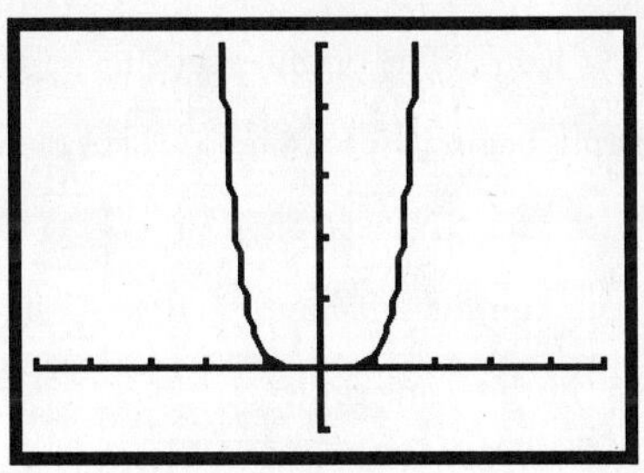

A table will also confirm this. Scroll through a table of values for y_1 and y_2 and observe that y_1 and y_2 appear to have the same value for a given value of x. Again, although the y-values could differ for an x-value that was not observed, the equation appears to be an identity.

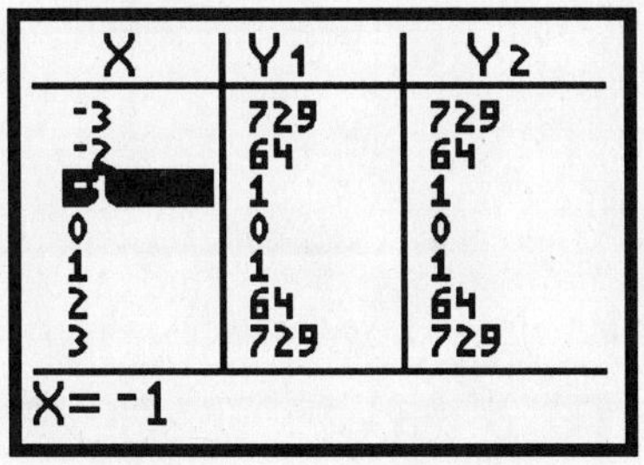

Example 9 (b), page 11: Determine whether $\sqrt{x+4} = \sqrt{x}+2$ appears to be an identity.

To determine whether this equation appears to be an identity, graph $y_1 = \sqrt{x+4}$ and $y_2 = \sqrt{x}+2$. Note that parentheses must be used on the TI-82 when entering $\sqrt{x+4}$: [2nd] [$\sqrt{\ }$] [(] [X, T, Θ] [+] 4 [)]. Without parentheses the expression entered would be $\sqrt{x}+4$. (Although the right parenthesis is optional, we include it for completeness.) The TI-83 forces the use of parentheses by producing "$\sqrt{\ }($" when [2nd] [$\sqrt{\ }$] is pressed and assumes a right parenthesis at the end of the expression if none is entered earlier. Thus, although it is not necessary to type parentheses when entering $\sqrt{x+4}$ on the TI-83, it is necessary to enter a right parenthesis in $y_2 = \sqrt{x} + 2$. Press [2nd] [$\sqrt{\ }$] [X, T, Θ] [)] [+] 2. If the right parenthesis is omitted, the expression becomes $\sqrt{x+2}$ rather than $\sqrt{x} + 2$.

Any window that includes a portion of the first quadrant will show that the graphs differ. Thus, the equation is not an identity.

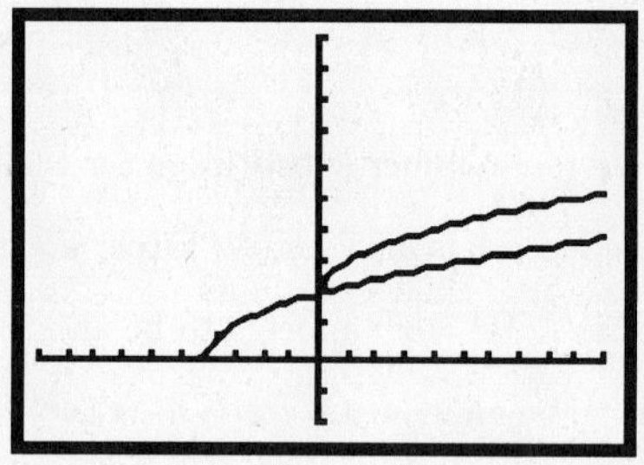

On the TI-83 different graph styles can be selected to allow us to differentiate visually between two or more graphs. Here, for example, we could have selected a solid line for y_1 and a dotted or thick line for y_2. Page 3-9 of the TI-83 Guidebook illustrates the available graph styles and their icons. The original settings of the TI-83 call for all equations to be graphed with a solid line. To change the style of a graph, begin by pressing [Y =] to display the "Y =" screen. To graph $y_2 = \sqrt{x}+2$ using a thick line, for example, we would first press [Y =]. Then press [▽] to move the cursor to y_2. Now press [◁] [◁] to move the cursor to the graph style icon in the far left-hand column beside y_2. Press [ENTER] repeatedly to rotate through the graph styles. These styles rotate in the same order in which they appear in the table on page 3-9 of the TI-83 Guidebook. When the thick line icon appears beside y_2, press [GRAPH] to display the graphs of y_1 and y_2.

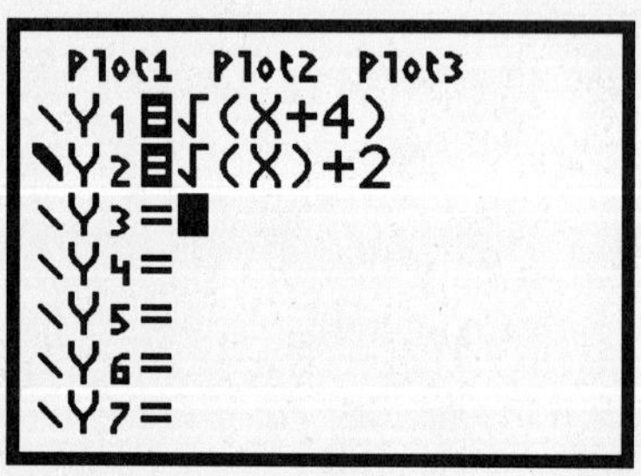

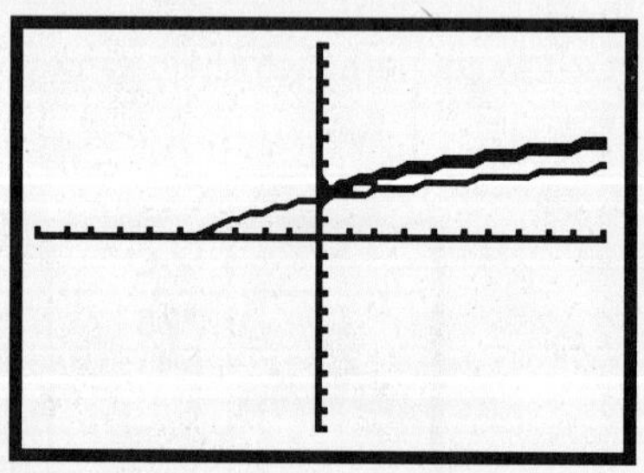

A table will also show that y_1 and y_2 do not always have the same value for a given x-value.

X	Y1	Y2
-2	1.4142	ERROR
-1	1.7321	ERROR
0	2	2
1	2.2361	3
2	2.4495	3.4142
3	2.6458	3.7321
4	2.8284	4

X=1

Note that y_1 and y_2 have the same value for $x = 0$ but not for the other possible substitutions shown. (The ERROR entries in the Y_2-column show that $x = -2$ and $x = -1$ cannot be substituted in $\sqrt{x} + 2$ to obtain a real number.)

Both the graph and the table demonstrate that the equation is not an identity.

SOLVING EQUATIONS USING TRACE AND ZOOM

There are several techniques that can be used to solve equations with a grapher. One uses the grapher's TRACE and ZOOM features.

Example 10, page 13: Solve $x^3 - 3x + 1 = 0$. Approximate the solutions to three decimal places.

The solutions of this equation are the first coordinates of the x-intercepts of the graph of $y = x^3 - 3x + 1$. To find these coordinates we first graph $y = x^3 - 3x + 1$ in a viewing window that shows all of the x-intercepts. Here we use the standard window, $[-10, 10, -10, 10]$.

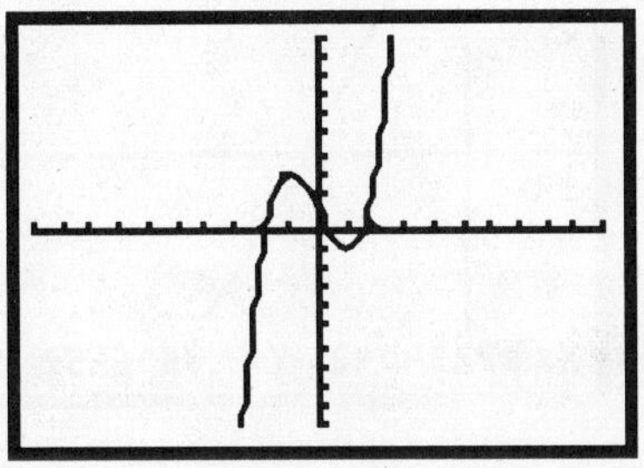

We see that x-intercepts occur near $x = -2$, $x = 0$, and $x = 2$. A portion of the viewing window can be enlarged near each of these values in order to find the desired three decimal place approximation. For example, let's examine the graph near $x = 0$.

Press TRACE. The TRACE cursor appears on the graph at the middle x-value of the window, in this case at $x = 0$. On the TI-82 the number 1 appears in the upper right-hand corner of the screen indicating that the cursor is on the graph of equation y_1. The equation of the curve being traced appears at the top of the screen on the TI-83. The x-and y-values at the bottom of the screen indicate the coordinates of the point where the cursor is positioned, in this case at $x = 0$, $y = 1$.

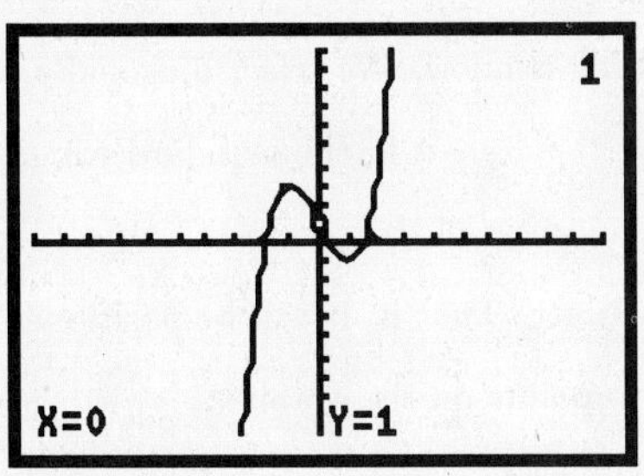

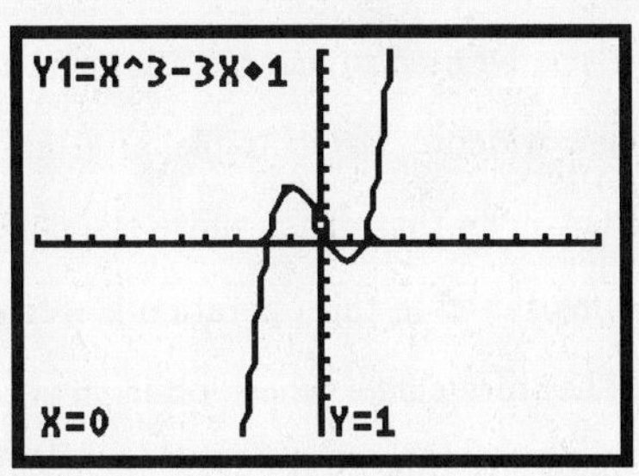

Pressing ◁ or ▷ moves the cursor to the left or right along the curve. Note that the TRACE cursor always remains on the curve.

In order to find the middle x-intercept, we enlarge the portion of the graph near $x = 0$ by first positioning the cursor as close to this intercept as possible. That is, near $x = 0$, position the cursor as close as possible to $y = 0$. Then press ZOOM 2 ENTER to zoom in on the graph near $x = 0$.

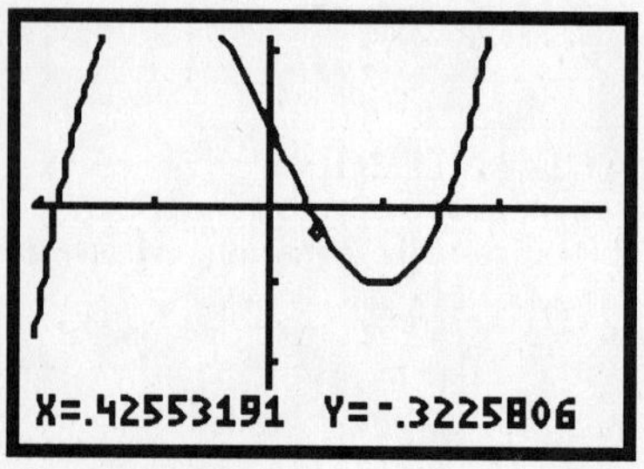

Press TRACE again and move the cursor as close as possible to $y = 0$ near $x = 0$. Now press ZOOM 2 ENTER again to enlarge this portion of the graph further.

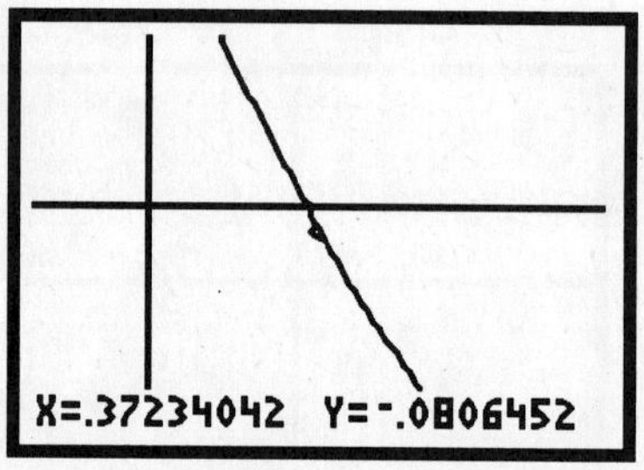

Continue tracing and zooming in until the x-values determined by positioning the cursor just to the left and just to the right of the x-intercept are the same when rounded to three decimal places. We find that $x \approx 0.347$ at the middle x-intercept. This is one solution of the equation $x^3 - 3x + 1 = 0$.

To find a second solution of the equation press ZOOM 6 to return to the standard window; then trace to a position on the curve near another x-intercept and zoom in as described above. Continue to trace and zoom in until the desired accuracy is obtained. Repeat this process to find the third solution. The other two solutions are about -1.879 and 1.532.

At any point in this process, we can zoom out to the previous window by pressing ZOOM 3 ENTER. This can be repeated as many times as desired.

The ZOOM BOX operation can also be used to enlarge a portion of the viewing window. We select diagonally opposed corners of a box that defines the new window. For example, after graphing $y = x^3 - 3x + 1$ in the standard window press ZOOM 1 to access the ZOOM BOX operation. Then move the cursor from the center of the screen to any corner of the box to be defined. Note that the cursor used in this operation is a free-moving cursor. That is, it can be positioned at any point on the screen as opposed to the trace cursor whose position is restricted to points on the graph.

Press ENTER and a small square dot appears, indicating that the first corner has been selected. Now use ◁, ▷, △, and/or ▽ to move the cursor to the corner of the box diagonally opposite this one. As you move the cursor away from the first corner the boundaries of the box appear on the screen and change as the position of the cursor changes.

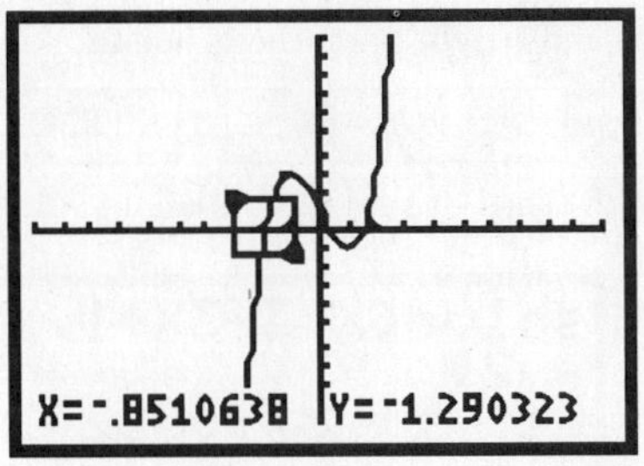

When the box is defined as you want it, press ENTER to show only the portion of the graph in the box.

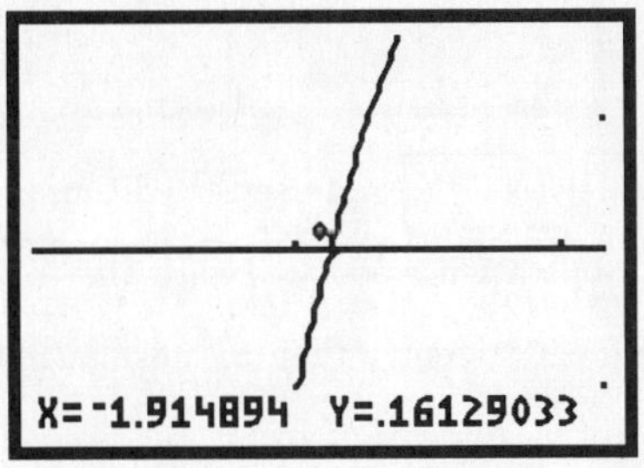

Now we can zoom in or use ZOOM BOX again to further enlarge a portion of the graph.

SOLVING EQUATIONS USING SOLVE OR SOLVER

The SOLVE operation on both the TI-82 and the TI-83 or the SOLVER operation on the TI-83 can also be used to solve equations. It is important to note that an equation must be written with 0 on one side of the equals sign when this operation is used.

Example 10, page 13: Solve $x^3 - 3x + 1 = 0$. Approximate the solutions to three decimal places.

Begin as in the procedure described above for using TRACE and ZOOM by graphing $y = x^3 - 3x + 1$ in a window that shows all of the x-intercepts. Note again that the x-intercepts occur near $x = -2$, $x = 0$, and $x = 2$. We will first find the solution near $x = -2$.

The TI-82's SOLVE operation is item 0 on the MATH menu. It follows item 9 and is not immediately visible when the MATH menu is selected. Access it by pressing MATH 0 or by pressing MATH, then using ▽ to scroll down and position the cursor over "0:" and finally pressing ENTER. The TI-83's SOLVE operation must be accessed through the CATALOG, an alphabetical list of all functions and instructions available on the TI-83. Select the CATALOG by pressing 2nd CATALOG. (CATALOG is the second operation associated with the 0 key.) Press ▽ to scroll down until the triangular selection cursor in the left-hand column is beside "solve(." Now press ENTER. To find SOLVE more quickly in the CATALOG, press S (by pressing the LN key) after 2nd CATALOG to skip to the first item beginning with the letter S. Then use ▽ to scroll down to "solve(."

On either grapher, the notation "solve(" appears on the screen with the blinking cursor positioned to the right of the parenthesis. We enter the left-hand side of the equation, the variable used in the equation, and a guess for the solution.

These entries must be separated by commas and a right parenthesis must follow the guess. Since we are seeking the solution near $x = -2$, we use this for the guess. Press [X, T, Θ] [∧] 3 [−] 3 [X, T, Θ] [+] 1 [,] [X, T, Θ] [,] [(−)] 2 [)] [ENTER]. The grapher returns the value −1.879385242. Thus one solution is about −1.879.

```
solve(X^3-3X+1,X
,-2)
       -1.879385242
```

To find the solution near $x = 0$ press [2nd] [ENTRY] and use the editing techniques described on page 6 of this manual to change −2 to 0. (ENTRY is the second operation associated with the [ENTER] key. It is used to recall the previous entry. Often it is more efficient to recall and edit an entry than to make an entirely new entry.) Then press [ENTER] and the value .3472963553 is returned. Instead of recalling and editing the previous entry, we could have selected the SOLVE operation from the MATH menu again and reentered the left-hand side of the equation, the variable, and the new guess. To find the solution near $x = 2$, repeat the previous process, changing 0 to 2. Thus, as we saw before, the other two solutions are about 0.347 and 1.532.

```
solve(X^3-3X+1,X
,0)
        .3472963553
solve(X^3-3X+1,X
,2)
        1.532088886
```

Since $y_1 = x^3 - 3x + 1$ has been entered in the "Y =" screen, we can enter Y_1 instead of $x^3 - 3x + 1$ in the SOLVE operation. To do this on the TI-82 press [2nd] [Y-VARS] 1 1 instead of [X, T, Θ] [∧] 3 [−] 3 [X, T, Θ] [+] 1. On the TI-83, press [VARS] [▷] 1 1.

```
solve(Y1,X,-2)
       -1.879385242
solve(Y1,X,0)
        .3472963553
solve(Y1,X,2)
        1.532088886
```

The TI-83's SOLVER operation is item 0 on the MATH menu. It follows item 9 and is not immediately visible when the MATH menu is selected. Access it by pressing [MATH] 0 or by pressing [MATH], then using [▽] to scroll down and position the cursor over "0:" and finally pressing ENTER. When this is done the EQUATION SOLVER screen is displayed. If SOLVER has previously been used, it will be necessary to press △ after MATH 0 to display the EQUATION SOLVER screen. Then press CLEAR to delete the previously entered equation. The cursor appears beside "eqn: 0 =."

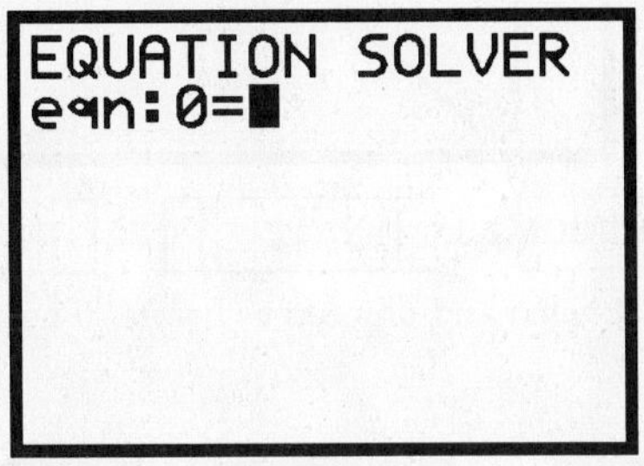

We can type $x^3 - 3x + 1$ followed by ENTER or, since $y_1 = x^3 - 3x + 1$ has been entered in the "Y =" screen, we can enter Y_1 by pressing VARS ▷ 1 1 ENTER. Now the interactive solver editor is displayed. To find the solution of the equation near $x = -2$, type the initial guess (−) 2 beside "X =." It is not necessary to change the "bound =" setting, although since we know that all of the solutions of this equation are between -3 and 3 we could move the cursor to this line and edit the entry there to $\{-3, 3\}$. This is optional, but it could help find the solution more quickly. Move the cursor to the line on which the variable X appears and press ALPHA SOLVE. (SOLVE is the Alpha operation associated with the ENTER key.) The solution X = −1.879385241... is displayed, so $x \approx -1.879$.

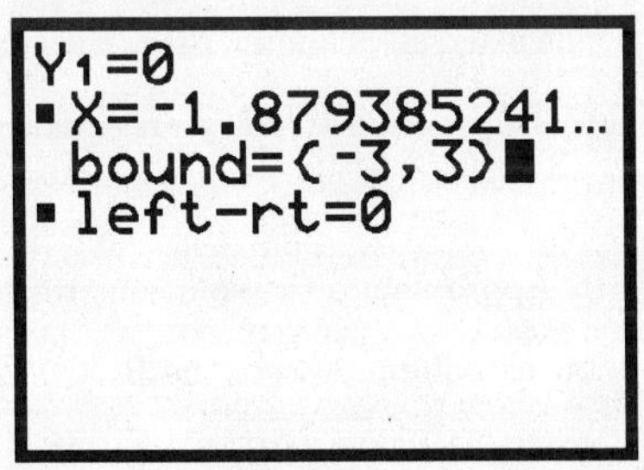

To find the solution near $x = 0$, simply edit X = −1.879385241... to X = 0 on the screen in which the first solution appeared, followed by ALPHA SOLVE. Find the solution near $x = 2$ similarly. We see that the other two solutions are about 0.347 and 1.532.

Example 12, page 15: Solve: $\frac{2}{3}x - 7 = 5$.

Since the SOLVE and SOLVER operations require an equation to be written with 0 on one side of the equals sign, we write this equation as $\frac{2}{3}x - 7 - 5 = 0$, or $\frac{2}{3}x - 12 = 0$. Then graph $y = \frac{2}{3}x - 12$ in a window that shows the x-intercept. Here we use $[-5, 35, -10, 10]$, Xscl = 5, Yscl = 2. Observe that the x-intercept is near $x = 20$. This is the value we will use as a guess for the solution.

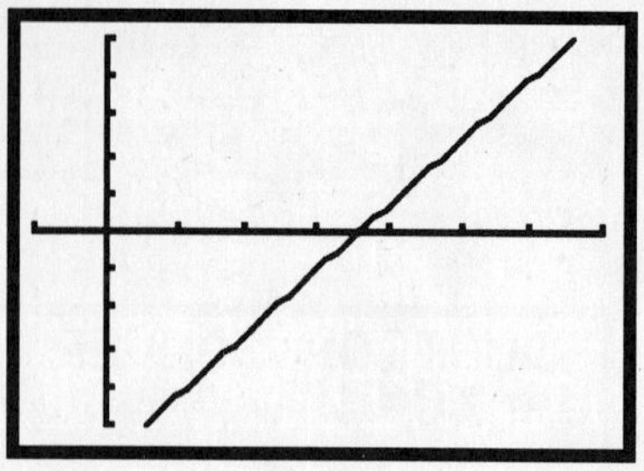

On the TI-82 press MATH 0 2nd Y-VARS 1 1 , X, T, Θ , 20) ENTER. On the TI-83, select SOLVE from the CATALOG or SOLVER from the MATH menu and proceed as before. The grapher returns the solution of the equation, 18.

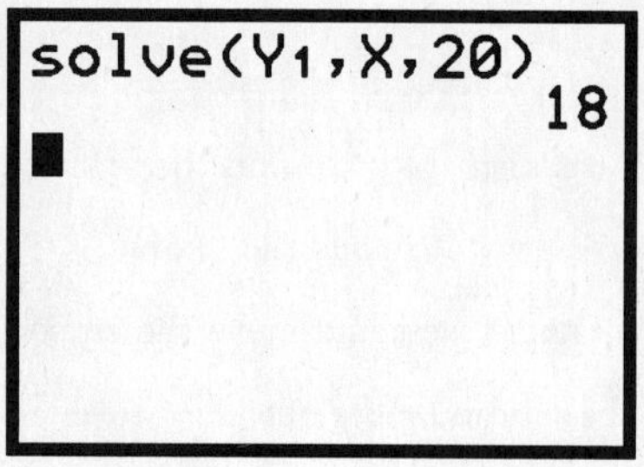

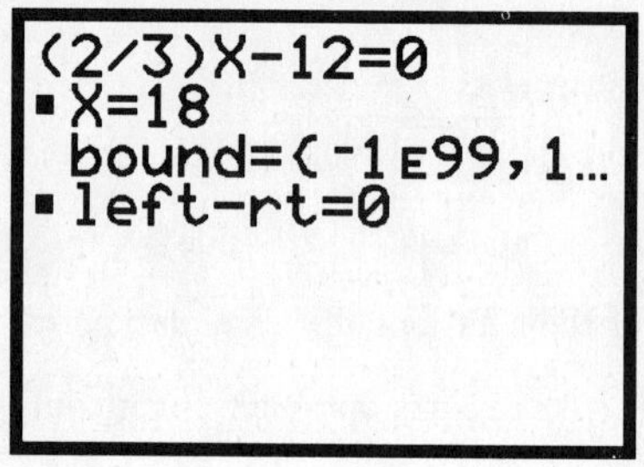

SOLVING EQUATIONS USING ROOT OR ZERO

A solution of an equation with zero on one side is usually called a zero, but is sometimes called a root. The TI-82's ROOT operation or the TI-83's ZERO operation can also be used to solve an equation. (These are actually the same operation but with different names.)

Example 10, page 13: Solve $x^3 - 3x + 1 = 0$. Approximate the solutions to three decimal places.

Graph $y = x^3 - 3x + 1$ in the standard window as before. Access the ROOT/ZERO operation by pressing 2nd CALC 2. (CALC is the second operation associated with the TRACE key.) The grapher prompts us to select, in succession, lower and upper bounds as well as a guess for the zero. We will illustrate by using ROOT/ZERO to find the solution of $x^3 - 3x + 1 = 0$ near $x = 0$.

After 2nd CALC 2 is pressed, the prompt "Lower Bound?" appears on the TI-82 screen below the graph. The TI-83 displays the prompt "Left Bound?" in the same location. Position the cursor just to the left of the x-intercept near 0 and press ENTER. The TI-82 prompt "Upper Bound?" or the TI-83 prompt "Right Bound?" then appears. Position the cursor just to the right of the intercept and press ENTER again. Now the prompt "Guess?" appears. Move the cursor as close to the x-intercept as possible and press ENTER. On the TI-83 we can also use the keypad to enter numbers for the left and right bounds and for the guess instead of using the cursor. Here we might enter 0, 1, and 0.5, respectively. The x-value that the grapher returns is the solution. We see again that the solution is about 0.347.

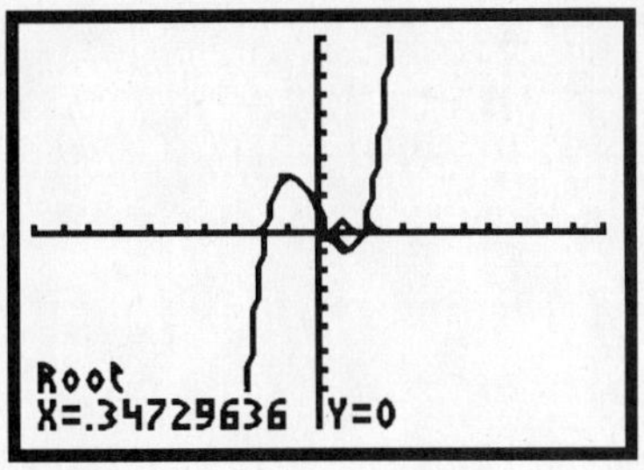

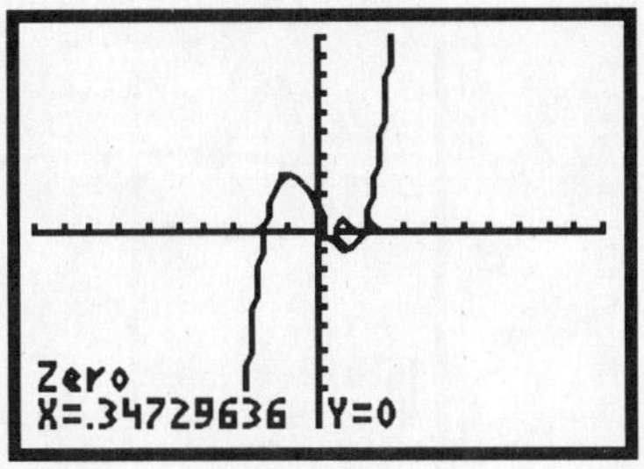

The other two solutions of the equation can be found by using the ROOT/ZERO operation two more times.

FINDING POINTS OF INTERSECTION

There are several ways in which the grapher can be used to determine the point(s) of intersection of two graphs.

Example 11, page 14: Find the point of intersection of the graphs of the equations $y_1 = 3x^5 - 20x^3$ and $y_2 = 34.7 - 1.28x^2$. Approximate the coordinates to three decimal places.

First graph the equations in a window that shows all of the points of intersection. We use $[-4, 4, -80, 80]$, Xscl = 1, Yscl = 20 here.

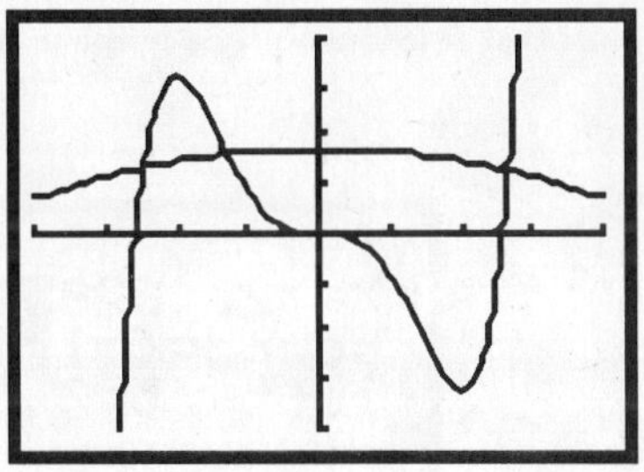

There are three points of intersection. One way to find their coordinates is to use TRACE and ZOOM. (See "Solving Equations Using Trace and Zoom" on page 13 of this manual.) Another method is to use the INTERSECT operation. We will illustrate its use by finding the point of intersection at the far left. After graphing y_1 and y_2 as above, access INTERSECT by pressing [2nd] [CALC] 5. The prompt "First curve?" appears on the screen below the graph. Move the cursor to the graph of y_1 and press [ENTER]. (If the cursor is positioned on the graph of y_1 when the prompt appears, simply press [ENTER].) Now the prompt "Second curve?" appears. Position the cursor on the graph of y_2 and press [ENTER]. The prompt "Guess?" then appears. Move the cursor as close as possible to the point of intersection and press [ENTER] again. On the TI-83 we can also use the keypad to enter a number for the guess, say −2.5. The grapher returns the coordinates of the point of intersection. We see that they are about $(-2.463, 26.936)$.

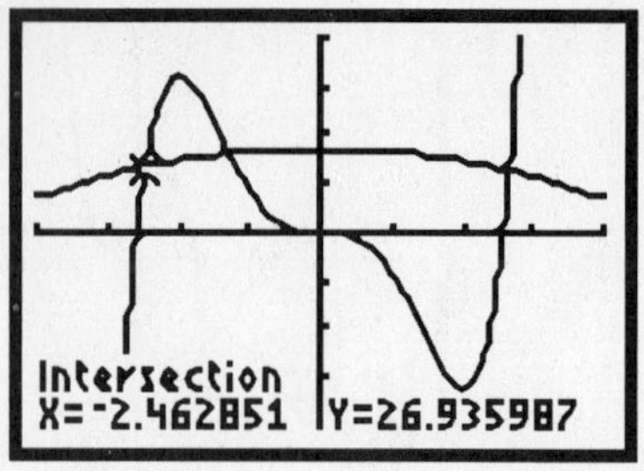

The coordinates of the other two points of intersection can be found by using the INTERSECT operation two more times.

The SOLVE or SOLVER operation can also be used to find the x-coordinates of the points of intersection by finding solutions of $y_1 = y_2$, or $y_1 - y_2 = 0$. Since the SOLVE and SOLVER operations require an equation to be written with 0 on one side of the equals sign, we must use the form $y_1 - y_2 = 0$.

Again it is necessary to graph y_1 and y_2 in order to find guesses for the solutions. To find the point of intersection on the far left, first note that its x-coordinate is about -2.5. This is the value we will use for the guess. On the TI-82 press [MATH] 0 [2nd] [Y-VARS] 1 1 [−] [2nd] [Y-VARS] 1 2 [,] [X, T, Θ] [,] [(−)] 2 [.] 5 [)] [ENTER]. Either SOLVE or SOLVER could be used on the TI-83 as described previously. The grapher returns the value -2.462851032. This is the first coordinate of the point of intersection.

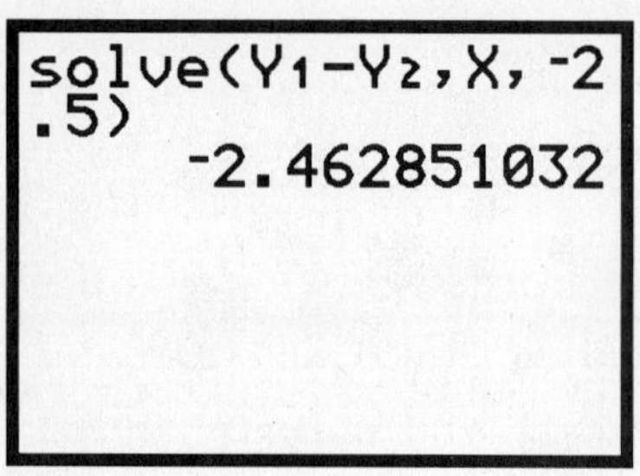

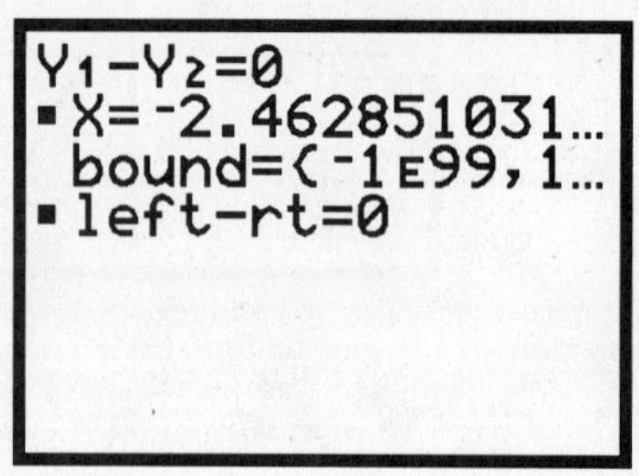

This value must be substituted into either y_1 or y_2 to find the second coordinate. To substitute in y_1 after using SOLVE press 3 [2nd] [ANS] [∧] 5 [−] 20 [2nd] [ANS] [∧] 3 [ENTER]. (ANS is the second operation associated with the [(−)] key. This operation recalls the answer to the previous computation, thus allowing it to be used in a new computation.) To substitute in y_1 after using SOLVER on the TI-83 first exit the SOLVER screen by pressing [2nd] [QUIT]. Then press 3 [X, T, Θ, n] [∧] 5 [−] 2 0 [X, T, Θ, n] [∧] 3 [ENTER]. The value 26.93598694 is returned.

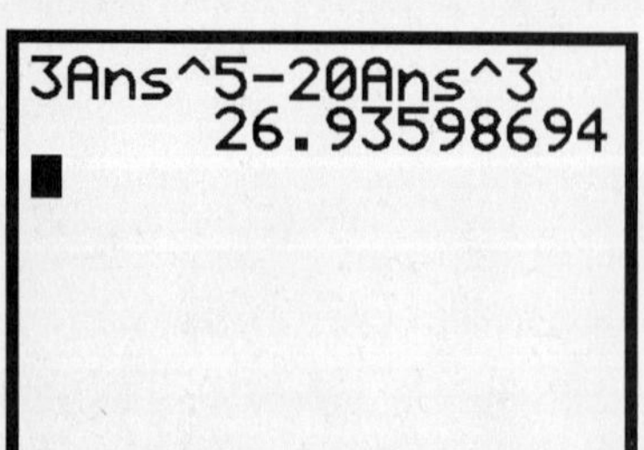

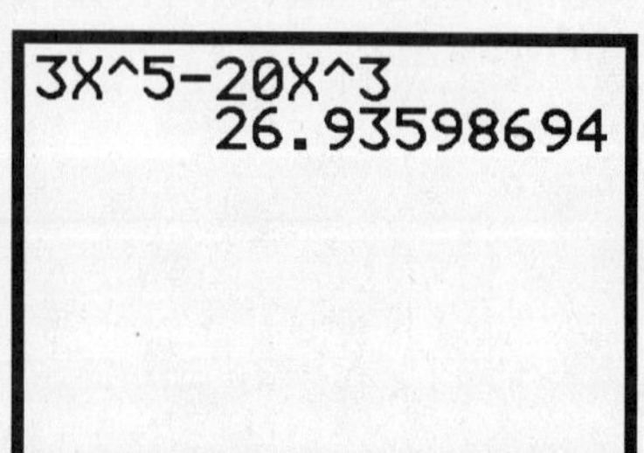

Thus, we see that the coordinates of the point of intersection are about $(-2.463, 26.936)$. If we had used the rounded x-value of -2.463 in y_1 to compute the corresponding y-value, the result would have been 26.908. If we had used the rounded x-value in y_2, the resulting y-value would have been 26.935. To achieve the greatest accuracy, then, it is important to use the unrounded x-value in this computation.

A table set in ASK mode can be used to find the second coordinate more quickly. After performing the SOLVE operation, press [2nd] [TABLE] [2nd] [ANS] [ENTER]. After using SOLVER, simply press [2nd] [TABLE] [X, T, Θ, n] [ENTER]. The value 26.936 appears in both Y-columns.

X	Y1	Y2
-2.463	26.936	26.936

X=

Note that, although the table shows the rounded x-value of -2.463, the unrounded x-value resulting from the SOLVE or SOLVER operation was actually used to compute the y-value. Confirm this by entering the rounded value -2.463 in the table. The table shows different values in the Y_1- and Y_2-columns, demonstrating that a less accurate result is produced with the rounded number and that the y-value computed depends on which equation is used.

X	Y1	Y2
-2.463	26.908	26.935

X=

SOLVING EQUATIONS USING INTERSECT

The INTERSECT operation can also be used to solve equations.

Example 12, page 15: Solve: $\frac{2}{3}x - 7 = 5$.

The solution of this equation is the first coordinate of the point of intersection of the graphs of $y_1 = \frac{2}{3}x - 7$ and $y_2 = 5$. Thus, we graph these equations in a window that shows the point of intersection and use INTERSECT as described in "Finding Points of Intersection," page 19 of this manual. The x-coordinate of the point of intersection is 18. This is the solution of the equation.

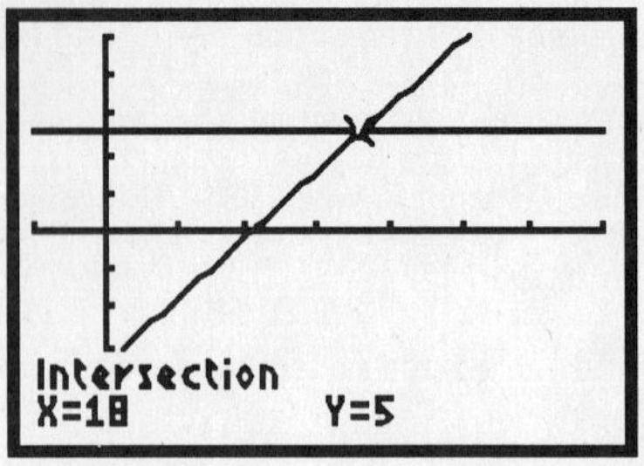

SELECTING DOT MODE

When graphing an equation in which a variable appears in a denominator, DOT mode should be used. To select DOT mode press MODE and use ▽ and ▷ to position the blinking cursor over "Dot."" Then press ENTER.

See page 16 in the text for further explanation of why DOT mode is used in certain cases and for an illustration of an equation graphed in both CONNECTED and DOT modes.

Chapter 1 Graphs, Functions, and Models

FINDING FUNCTION VALUES

When a formula for a function is given, function values for real-numbered inputs can be found in several ways on a grapher.

Example 4 (a), (b), page 23 (Page numbers refer to pages in the text.): For $f(x) = 2x^2 - x + 3$, find $f(0)$ and $f(-7)$.

Method 1: Substitute the inputs directly in the formula. For example, to find $f(0)$ press 2 [×] 0 [x^2] [−] 0 [+] 3 [ENTER]. To find $f(-7)$ press 2 [(] [(−)] 7 [)] [x^2] [−] [(] [(−)] 7 [)] [+] 3 [ENTER].

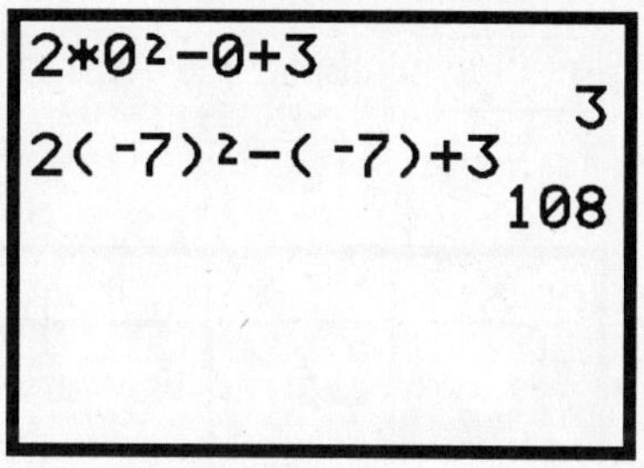

Method 2: Enter $y_1 = 2x^2 - x + 3$ in the "Y =" screen. Then press [2nd] [QUIT] to go to the home screen. (QUIT is the second operation associated with the [MODE] key.) Now to use the TI-82 to find $f(0)$, the value of y_1 when $x = 0$, press 0 [STO ▷] [X, T, Θ] [2nd] [:] [2nd] [Y-VARS] 1 1 [ENTER]. (: is the second operation associated with the [.] key.) This series of keystrokes stores 0 as the value of x and then substitutes it in the function y_1. To find the function value at $x = -7$ on the TI-82, repeat this process, using −7 in place of 0 or press [2nd] [ENTRY] to recall the previous entry, move the cursor to 0, and replace 0 with −7. Note that since two keystrokes are required to enter −7, it is necessary to use [2nd] [INS] to change 0 to −7. That is, with the cursor on 0 press [(−)] [2nd] [INS] 7 [ENTER] to make this change.

To use the TI-83 to find $f(0)$ using this method press 0 [STO ▷] [X, T, Θ, n] [ALPHA] [:] [VARS] [▷] 1 1 [ENTER]. (: is the ALPHA operation associated with the [.] key.) To find $f(-7)$ either repeat this process using −7 in place of 0 or recall and edit the first entry as described above.

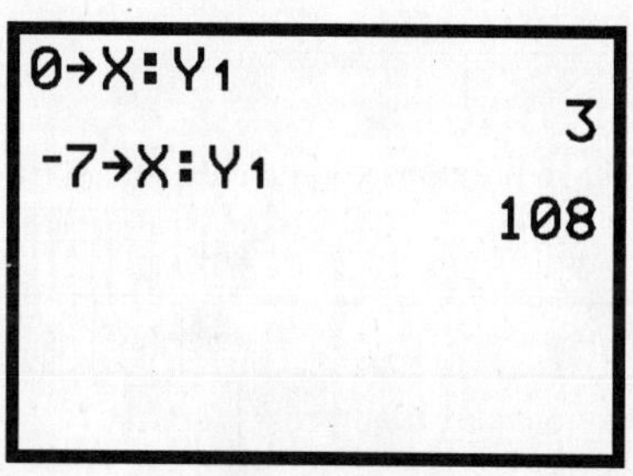

Method 3: Enter $y_1 = 2x^2 - x + 3$ in the "Y =" screen and press [2nd] [QUIT] to go to the home screen. Then, to find $f(0)$ on the the TI-82 press [2nd] [Y-VARS] 1 1 [(] [0] [)] [ENTER]. On the TI-83 press [VARS] [▷] 1 1 [(] [0] [)] [ENTER]. Note that these entries closely resemble function notation. To find $f(-7)$ either repeat this process entering -7 in place of 0 or recall the first entry and replace 0 with -7 as described in Method 2 above and then press [ENTER].

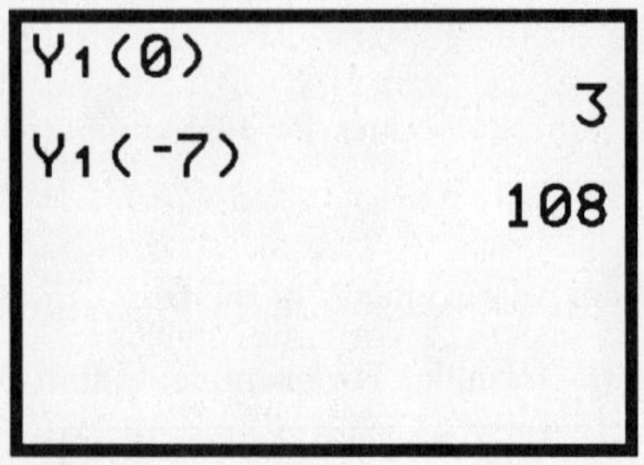

Method 4: The TABLE feature can also be used to find function values. With $y_1 = 2x^2 - x + 3$ entered in the "Y =" screen and the table set in Ask mode, press [2nd] [TABLE] to display the TABLE screen. Then press 0 [ENTER] to find $f(0)$ and [(−)] 7 [ENTER] to find $f(-7)$.

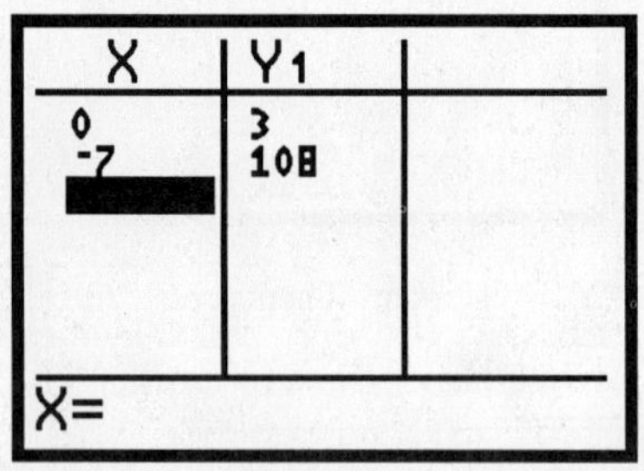

GRAPHING FUNCTIONS DEFINED PIECEWISE

Operations from the TEST menu are used to enter functions that are defined piecewise. (TEST is the second operation associated with the [MATH] key.) Select DOT mode from the MODE menu when graphing such functions. On the TI-83, DOT mode can also be selected by choosing the DOT style on the "Y =" screen. (See page 12 of this manual.) Any style selected on the TI-83 "Y =" screen overrides a MODE selection.

Example 4, page 38: Graph

$$f(x) = \begin{cases} 4, & \text{for } x \le 0, \\ 4 - x^2, & \text{for } 0 < x \le 2, \\ 2x - 6, & \text{for } x > 2. \end{cases}$$

Press [Y =] and clear any functions that have previously been entered. With the cursor beside "Y1 =" enter the function as described in the middle of page 102 of the text by pressing [(] [4] [)] [(] [X, T, Θ] [2nd] [TEST] 6 0 [)] [+] [(] [4] [−] [X, T, Θ] [x^2] [)] [(] 0 [2nd] [TEST] 5 [X, T, Θ] [)] [(] [X, T, Θ] [2nd] [TEST] 6 2 [)] [+] [(] [2] [X, T, Θ] [−] [6] [)] [(] [X, T, Θ] [2nd] [TEST] 3 2 [)]. Select a window and then press [GRAPH].

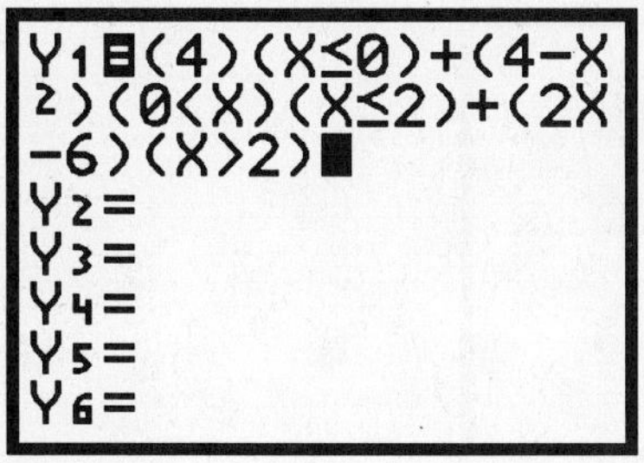

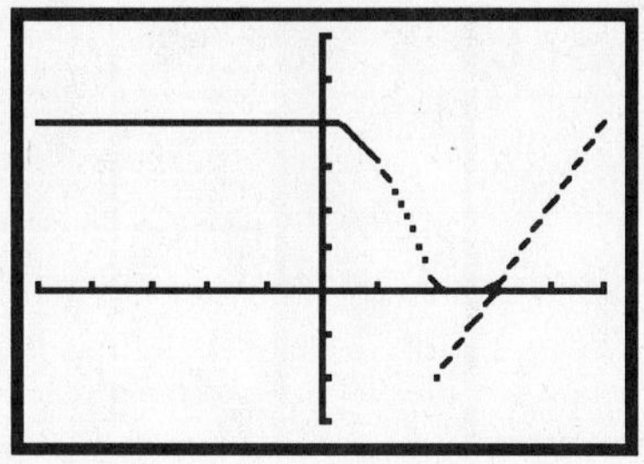

Example 6, page 39: Graph $f(x) = \text{INT}(x)$.

With the grapher set in DOT mode, press Y = and clear any previously entered functions. Position the cursor beside "Y1 =" and select the greatest integer function from the MATH NUM menu as follows. On the TI-82 press MATH ▷ 4 X, T, Θ and on the TI-83 press MATH ▷ 5 X, T, Θ, n). (The right parenthesis is optional.) Select a window and press GRAPH.

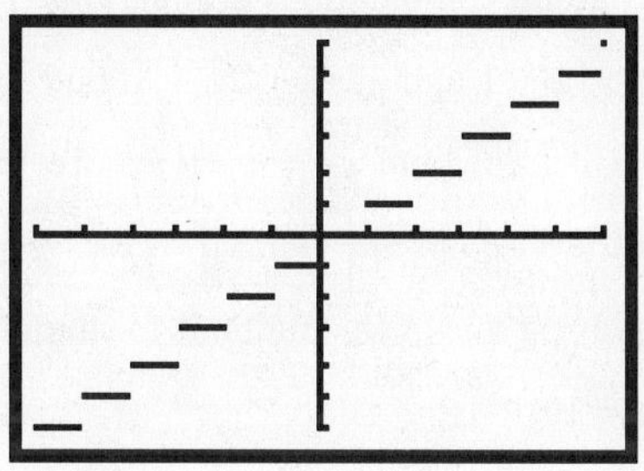

GRAPHING VERTICAL LINES

Although equations of vertical lines cannot be entered in the grapher in "Y =" form, vertical lines can be graphed using the DRAW feature. (DRAW is the second operation associated with the PRGM key.)

Interactive Discovery, top of page 56: Graph $x = -5$ and $x = 2$.

To graph $x = -5$, first press Y = and clear all previously entered equations. Then select a window. Now press 2nd QUIT to go to the home screen. Press 2nd DRAW 4 to select a vertical line. Then press (−) 5 ENTER to see the graph of $x = -5$. There are several ways to graph $x = -5$ and $x = 2$ on the same screen. For instance, after pressing 2nd DRAW 4 (−) 5 on the TI-82 press 2nd : 2nd DRAW 4 2 ENTER. On the TI-83 press ALPHA : rather than 2nd :. If you have pressed 2nd DRAW 4 (−) 5 followed by ENTER, press 2nd QUIT to leave the graph screen and then press 2nd DRAW 4 2 ENTER to add the graph of $x = 2$.

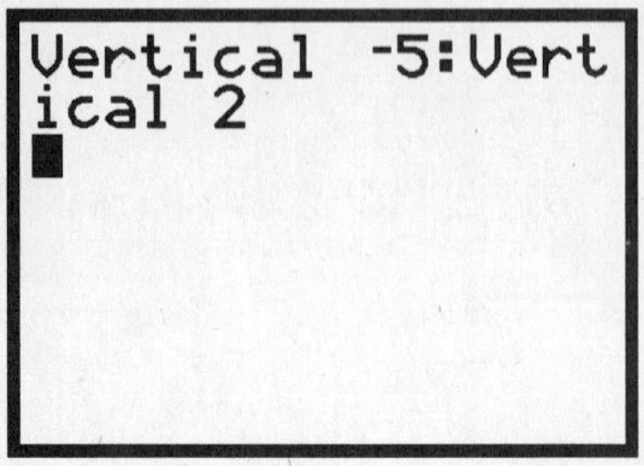

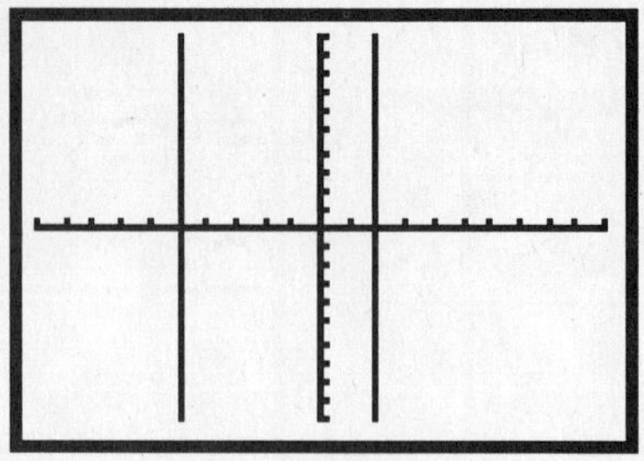

These drawings can be cleared from the graph screen by pressing [2nd] [DRAW] 1 to select the ClrDraw (clear drawing) operation. If ClrDraw was not accessed from the graph screen, it must be followed by [ENTER]. The drawings will also be cleared when another function is subsequently entered on the "Y =" screen and graphed.

THE ALGEBRA OF FUNCTIONS

The grapher can be used to evaluate and graph combinations of functions.

Example 1 (b), page 92: Given that $f(x) = x + 1$ and $g(x) = \sqrt{x+3}$, find $(f + g)(6)$.

Press [Y =] and enter $y_1 = x + 1$, $y_2 = \sqrt{x+3}$, and $y_3 = y_1 + y_2$. To enter $y_3 = y_1 + y_2$ on the TI-82 press [2nd] [Y-VARS] 1 1 [+] [2nd] [Y-VARS] 1 2. On the TI-83 press [VARS] [▷] 1 1 [+] [VARS] [▷] 1 2. Note that $y_3 = f(x)+g(x)$, or $(f+g)(x)$. Use y_3 to find $(f+g)(6)$ employing one of the methods for finding function values described on pages 23 and 24 of this manual. We find that $(f + g)(6) = 10$.

To view the graphs of $f(x)$, $g(x)$, and $(f + g)(x)$ enter y_1, y_2, and y_3 as above, select a window, and press [GRAPH]. These graphs appear on page 92 of the text. It is possible to deselect one or two of these functions and display the graph(s) of the remaining function(s). For example, to display only the graph of y_3 without deleting the equations of y_1 and y_2, press [Y =]. Then move the cursor to y_1, position it over the equals sign, and press [ENTER]. This deselects or turns off y_1. Do the same for y_2. Now press [GRAPH] and see only the graph of y_3.

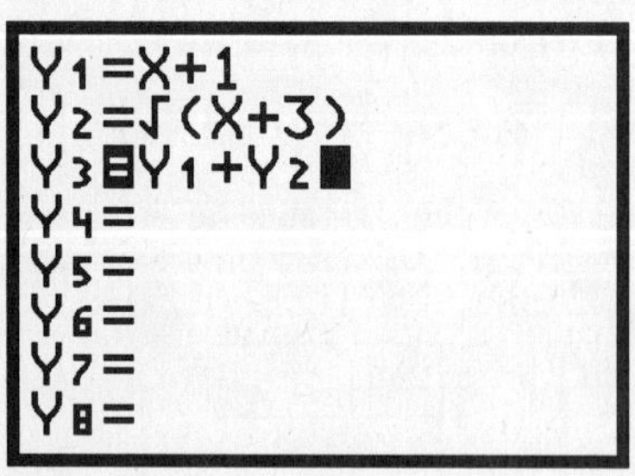

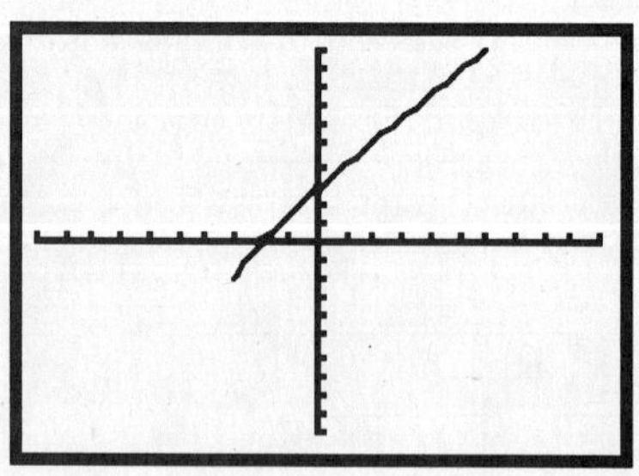

To select or turn on a function again, repeat this process. Note that the equals sign on a selected function is highlighted.

Example 3 (a), page 95: Given that $f(x) = 2x - 5$ and $g(x) = x^2 - 3x + 8$, find $(f \circ g)(7)$ and $(g \circ f)(7)$.

Press [Y =] and enter $y_1 = 2x - 5$, $y_2 = x^2 - 3x + 8$, $y_3 = 2y_2 - 5$, and $y_4 = y_1^2 - 3y_1 + 8$. Use the [Y-VARS] feature on a TI-82 or the [VARS] feature on a TI-83 as described above to enter y_3 and y_4. Note that $y_3 = (f \circ g)(x)$ and $y_4 = (g \circ f)(x)$. Use y_3 and y_4 to find $(f \circ g)(7)$ and $(g \circ f)(7)$, respectively, employing one of the methods for finding function values described on pages 23 and 24 of this manual.

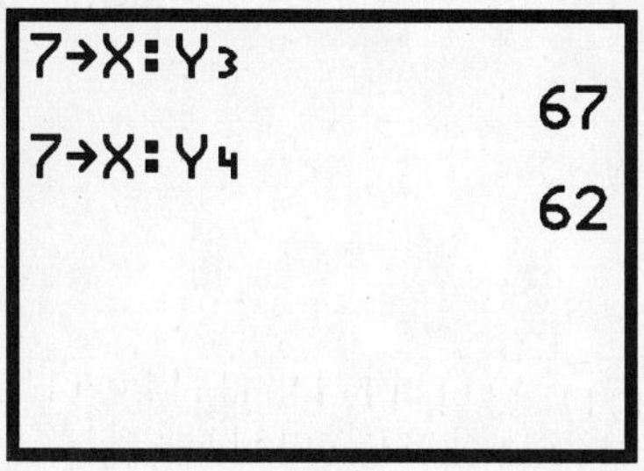

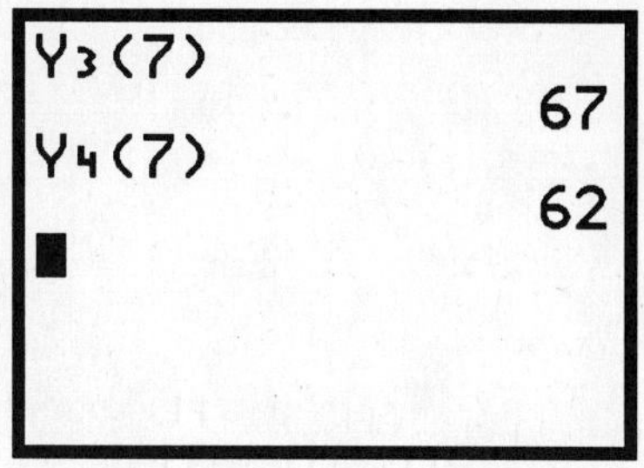

GRAPHING AN INVERSE FUNCTION

The DrawInv operation on the TI-82 and the TI-83 can be used to graph a function and its inverse on the same screen. A formula for the inverse function need not be found in order to do this. The grapher must be set in Func mode when this operation is used.

Example 7, page 105: Graph $f(x) = 2x - 3$ and $f^{-1}(x)$ using the same set of axes.

Enter $y_1 = 2x - 3$ and either clear or deselect all other functions on the "Y =" screen. Then press [2nd] [DRAW] 8 to select the DrawInv operation. (DRAW is the second operation associated with the [PRGM] key.) On the TI-82 follow these keystrokes with [2nd] [Y-VARS] 1 1 to select function y_1. On the TI-83 y_1 is selected by pressing [VARS] [▷] 1 1. Press [ENTER] to see the graph of the function and its inverse.

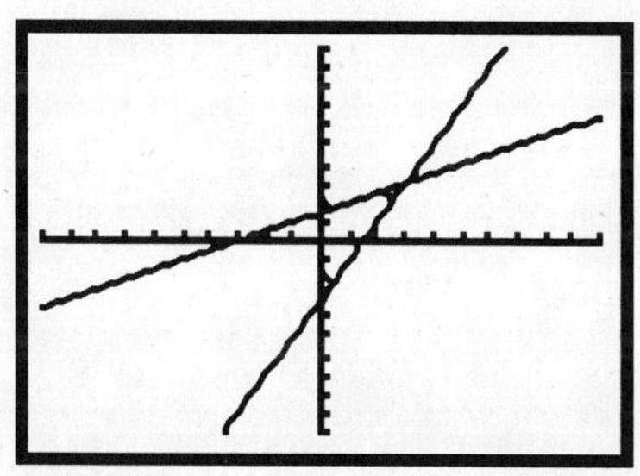

Chapter 2
The Trigonometric Functions

CONVERTING BETWEEN D°M′S″ AND DECIMAL DEGREE MEASURE

The ANGLE feature on the TI-82 and the TI-83 can be used to convert D°M′S″ notation to decimal notation and vice versa.

Example 5, page 122: Convert 5°42′30″ to decimal degree notation.

Enter 5°42′30″ as 5′42′30′ by pressing 5 [2nd] [ANGLE] 2 4 2 [2nd] [ANGLE] 2 3 0 [2nd] [ANGLE] 2 [ENTER]. (ANGLE is the second operation associated with the [MATRIX] key.) On the TI-83 enter 5°42′30″ by pressing 5 [2nd] [ANGLE] 1 4 2 [2nd] [ANGLE] 2 3 0 [ALPHA] [″] [ENTER]. (″ is the ALPHA operation associated with the [+] key.) In either case the grapher returns 5.708333333, so 5°42′30″ ≈ 5.71°.

Example 6, page 122: Convert 72.18°to D°M′S″ notation.

Press 7 2 [·] 1 8 [2nd] [ANGLE] 4 [ENTER]. The grapher returns 72°10′48″.

The conversions in Examples 5 and 6 are shown for the TI-82 on the left below and for the TI-83 on the right.

FINDING TRIGONOMETRIC FUNCTION VALUES

The grapher's SIN, COS, and TAN operations can be used to find the values of trigonometric functions. When angles are given in degree measure, the grapher must be set in Degree mode.

Example 7, page 122: Find the trigonometric function value, rounded to four decimal places, of each of the following.

a) tan 29.7° b) sec 48° c) sin 84°10′39″

a) Press [TAN] 2 9 [·] 7 [ENTER]. We find that tan 29.7° ≈ 0.5704.

b) The secant, cosecant, and cotangent functions can be found by taking the reciprocals of the cosine, sine, and tangent functions, respectively. This can be done either by entering the reciprocal or by using the [x^{-1}]. To find sec 48° we can enter the reciprocal of cos 48° by pressing 1 [÷] [COS] 4 8 [ENTER]. To find sec 48° using the [x^{-1}] key on the TI-82 press [(] [COS] 4 8 [)] [x^{-1}] [ENTER]. On the TI-83 press [COS] 4 8 [)] [x^{-1}] [ENTER]. The result is sec 48° ≈ 1.4945.

The output for the TI-82 is shown on the left below, and the TI-83 output is on the right.

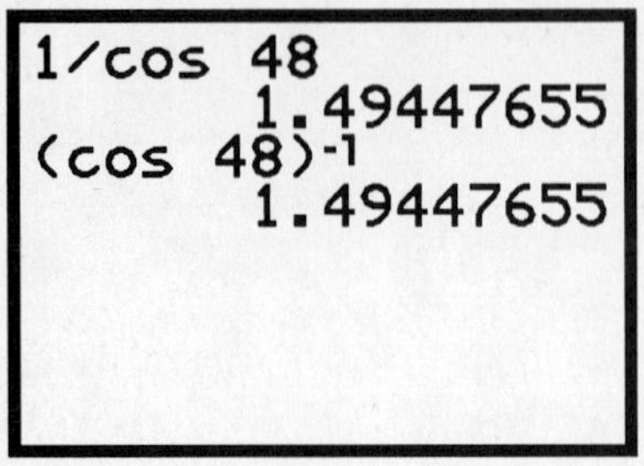

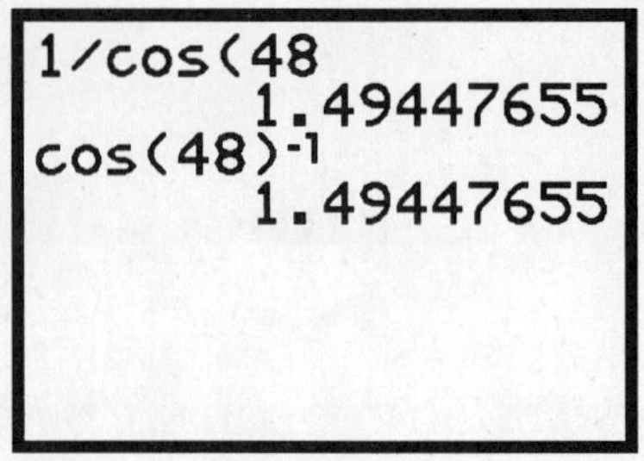

c) Press [SIN] followed by 84°10′39″ entered as described above in Converting Between D°M′S″ and Decimal Degree Measure for the grapher being used. Then press [ENTER]. We find that sin 84°10′39″ ≈ 0.9948.

FINDING ANGLES

The inverse trigonometric function keys provide a quick way to find an angle given a trigonometric function value for that angle.

Example 8, page 123: Find the acute angle, to the nearest tenth of a degree, whose sine value is approximately 0.20113.

Although the TABLE feature can be used to approximate this angle, it is faster to use the inverse sine key. With the grapher set in Degree mode press [2nd] [SIN^{-1}] [.] 2 0 1 1 3 [ENTER]. (SIN^{-1} is the second operation associated with the [SIN] key.) We find that the desired acute angle is approximately 11.6°.

Exercise 63, page 126: Find the acute angle, to the nearest tenth of a degree, whose cotangent value is 2.127.

Angles whose secant, cosecant, or cotangent values are known can be found using the reciprocals of the cosine, sine, and tangent functions, respectively. Since $\cot\theta = \dfrac{1}{\tan\theta} = 2.127$, we have $\tan\theta = \dfrac{1}{2.127}$, or $(2.127)^{-1}$. To find θ press [2nd] [TAN^{-1}] [(] 1 [÷] 2 [.] 1 2 7 [)] [ENTER] or [2nd] [TAN^{-1}] 2 [.] 1 2 7 [x^{-1}] [ENTER]. (TAN^{-1} is the second operation associated with the [TAN] key.) Note that the parentheses are necessary in the first set of keystrokes. The left parenthesis must be keyed in on the TI-82 while it appears on the TI-83 along with "$\tan^{-1}$." Without parentheses we would be finding the angle whose tangent is 1 and then dividing that angle by 2.127. We find that $\theta \approx 25.2°$.

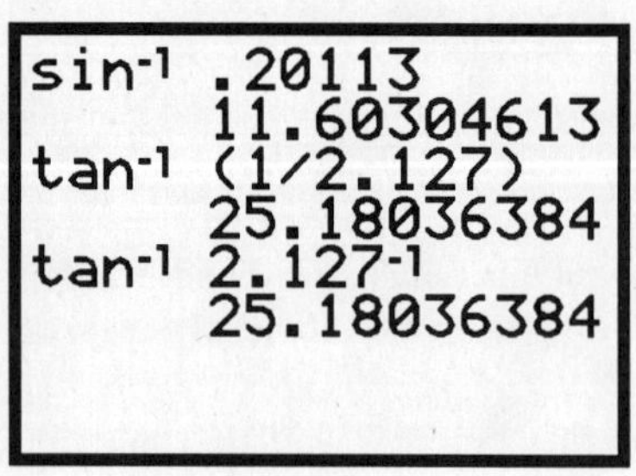

CONVERTING BETWEEN DEGREE AND RADIAN MEASURE

We can use the grapher to convert from degree to radian measure and vice versa. The grapher should be set in Radian mode when converting from degree to radian measure and in Degree mode when converting from radian to degree measure.

Example 3, page 157: Convert each of the following to radians.

a) 120° b) −297.25°

a) Set the grapher in Radian mode. Press 1 2 0 [2nd] [ANGLE] 1 [ENTER] to enter 120°. The grapher returns a decimal approximation of the radian measure. We see that 120° ≈ 2.09 radians.

b) With the grapher set in Radian mode press [(−)] 2 9 7 [.] 2 5 [2nd] [ANGLE] 1 [ENTER]. We see that −297.25° ≈ −5.19 radians.

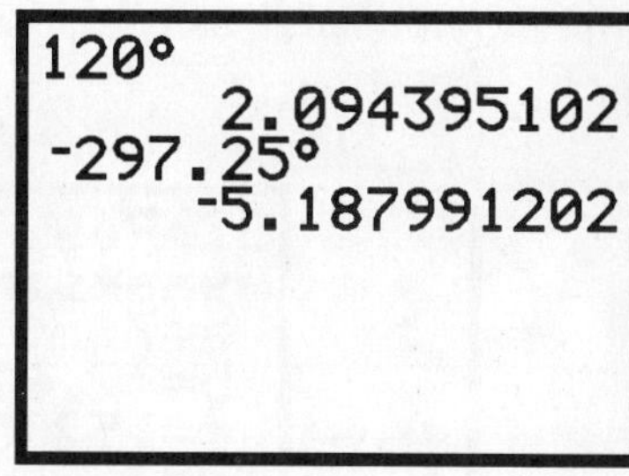

Example 4, page 157: Convert each of the following to degrees.

a) $\frac{3\pi}{4}$ radians b) 8.5 radians

a) Set the grapher in Degree mode. Then press [(] 3 [2nd] [π] [÷] 4 [)] [2nd] [ANGLE] 3 [ENTER] to enter $\frac{3\pi}{4}$ radians. (π is the second operation associated with the [∧] key). The grapher returns 135, so $3\pi/4$ radians = 135°. Note that the parentheses are necessary in order to enter the entire expression in radian measure. Without the parentheses, the grapher reads only the denominator, 4, in radian measure and an incorrect result occurs.

b) With the grapher set in Degree mode press 8 [.] 5 [2nd] [ANGLE] 3 [ENTER]. The grapher returns 487.0141259, so 8.5 radians ≈ 487.01°.

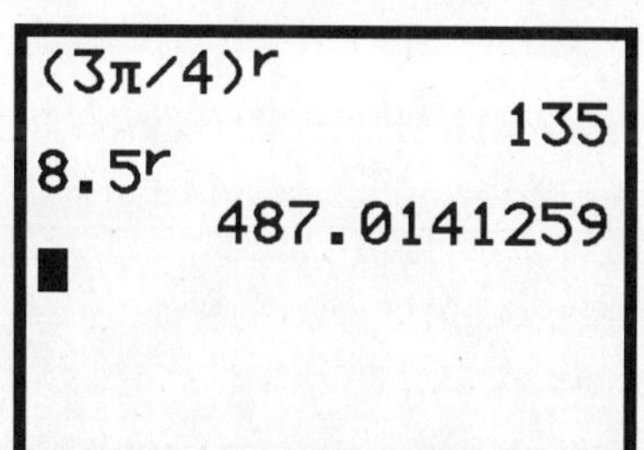

SINE REGRESSION

The SinReg operation on the TI-83 can be used to fit a sine curve $y = a\sin(bx + c) + d$ to a set of data. At least four data points are required and there must be at least two data points per period. The output of SinReg is always in radians, regardless of the Radian/Degree mode setting.

The operations of entering data, making scatterplots, and graphing and evaluating the regression function are the same as for linear regression functions. Read the section on Linear Regression and Scatterplots beginning on page 53 of this manual to review these procedures.

Exercise 57 (a), page 196: Fit a sine function to the given data on ski sales.

Enter the data in L_1 and L_2 as described in Appendix A.3 (page 53) of this manual. Press [STAT] [▷] to view the STAT CALC menu. Then select SinReg by pressing [ALPHA] C. If the data is entered in a combination of lists other than L_1 and L_2 their names must be entered separated by a comma. Since we have used L_1 and L_2 this is not necessary in this case. We also have the usual TI-83 option of specifying a $y =$ variable to which the regression equation can be stored. To select y_1, for example, press [VARS] [▷] 1 1. Now press [ENTER] to see the coefficients a, b, c, and d of the sine regression function $y = a\sin(bx + c) + d$

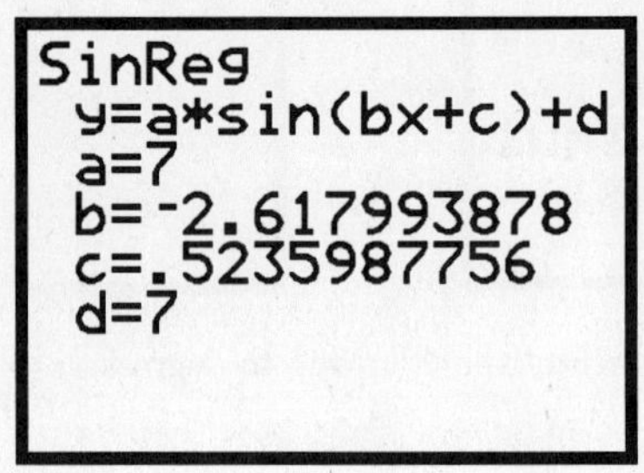

Chapter 3
Trigonometric Identities, Inverse Functions, and Equations

FINDING INVERSE FUNCTION VALUES

We can use a grapher to find inverse function values in both radians and degrees.

Example 2 (a), (e), page 235: Approximate $\cos^{-1}(-0.2689)$ and $\csc^{-1} 8.205$ in both radians and degrees.

To find inverse function values in radians, first set the grapher in Radian mode. Then, to approximate $\cos^{-1}(-0.2689)$, press [2nd] [COS^{-1}] [(−)] . 2 6 8 9 [ENTER]. The grapher returns 1.84304711, so $\cos^{-1}(-0.2689) \approx 1.8430$ radians.

To find $\csc^{-1} 8.205$, recall the identity $\csc\theta = \dfrac{1}{\sin\theta}$. Then $\csc^{-1} 8.205 = \sin^{-1}\left(\dfrac{1}{8.205}\right)$. On the TI-82 press [2nd] [SIN^{-1}] [(] 1 [÷] 8 . 2 0 5 [)] [ENTER] or [2nd] [SIN^{-1}] 8 . 2 0 5 [x^{-1}] [ENTER]. The readout is .1221806653, so $\csc^{-1} 8.205 \approx 0.1222$ radians. The TI-83 supplies the left parenthesis when [2nd] [SIN^{-1}] is pressed, so it is not necessary to enter this using a separate keystroke. The right parenthesis is optional on both the TI-82 and the TI-83.

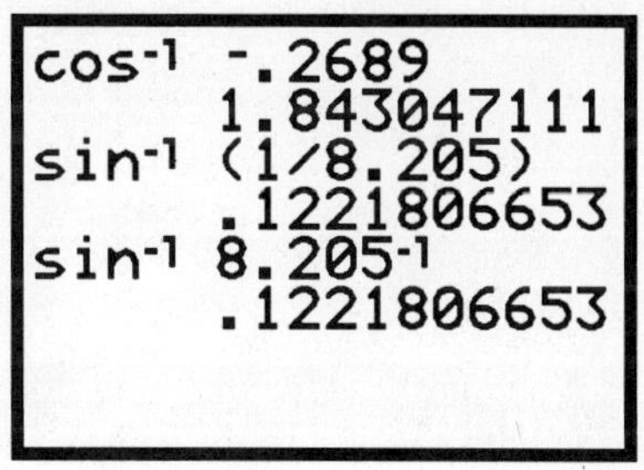

To find inverse function values in degrees, set the grapher in degree mode. Then use the keystrokes above to find that $\cos^{-1}(-0.2689) \approx 105.6°$ and $\csc^{-1} 8.205 \approx 7.0°$. We also use reciprocal relationships to find function values for arcsecant and arccotangent.

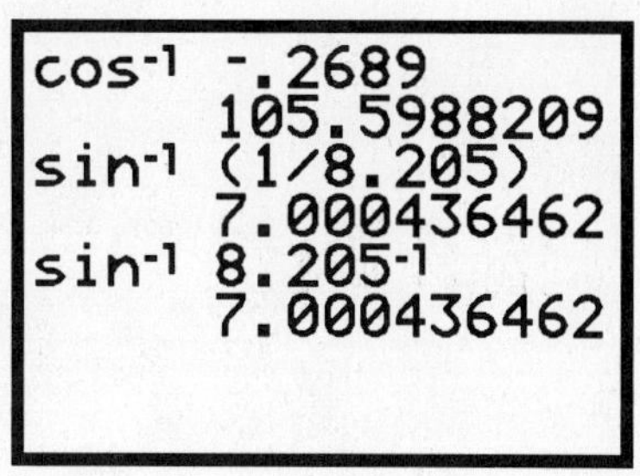

Chapter 4
Applications of Trigonometry

OPERATIONS WITH COMPLEX NUMBERS

Operations with complex numbers can be performed on the TI-83. First set the grapher in the complex $a + bi$ mode by pressing [MODE], positioning the cursor over $a + bi$, and pressing [ENTER].

Example 4, page 281:

(a) Add: $(8 + 6i) + (3 + 2i)$.

To find this sum press 8 [+] 6 [2nd] [i] [+] 3 [+] 2 [2nd] [i] [ENTER]. (The number i is the second operation associated with the [·] key.) Note that it is not necessary to include parentheses when we are adding.

(b) Subtract: $(4 + 5i) - (6 - 3i)$.

Press 4 [+] 5 [2nd] [i] [−] [(] 6 [−] 3 [2nd] [i] [)] [ENTER]. Note that, although the first set of parentheses is optional, the second set must be included so that the entire number $6 - 3i$ is subtracted.

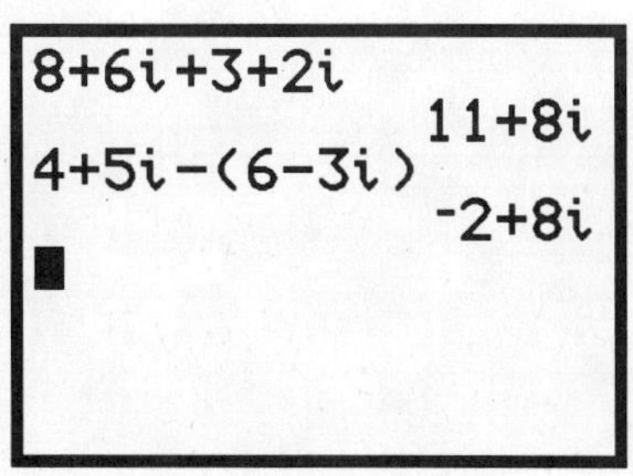

Example 5 (a), (b), (d), (e), page 282:

(a) Multiply: $\sqrt{-16} \cdot \sqrt{-25}$.

Press [2nd] [$\sqrt{\ }$] [(−)] 1 6 [)] [2nd] [$\sqrt{\ }$] [(−)] 2 5 [)] [ENTER].

(b) Multiply: $\sqrt{-5} \cdot \sqrt{-7}$.

Press [2nd] [$\sqrt{\ }$] [(−)] 5 [)] [2nd] [$\sqrt{\ }$] [(−)] 7 [)] [ENTER]. Note that this operation produces a decimal approximation of the product. That is, $\sqrt{-5} \cdot \sqrt{-7} \approx -5.916079783$. The exact value of the product is $-\sqrt{35}$.

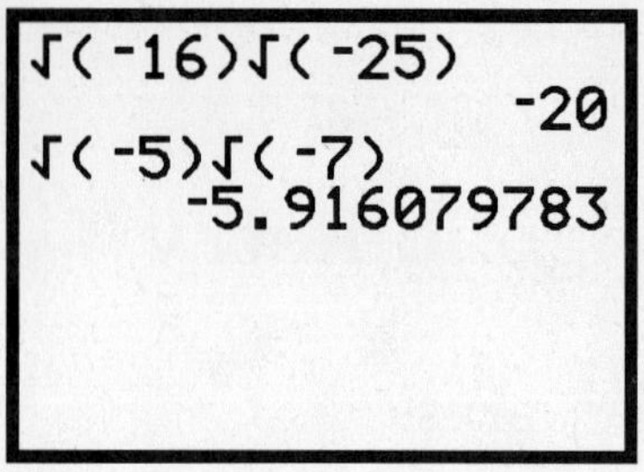

(d) Multiply: $(1 + 2i)(1 + 3i)$.

Press [(] 1 [+] 2 [2nd] [i] [)] [(] 1 [+] 3 [2nd] [i] [)] [ENTER].

(e) Multiply: $(3 - 7i)^2$.

Press [(] 3 [−] 7 [2nd] [i] [)] [x^2] [ENTER].

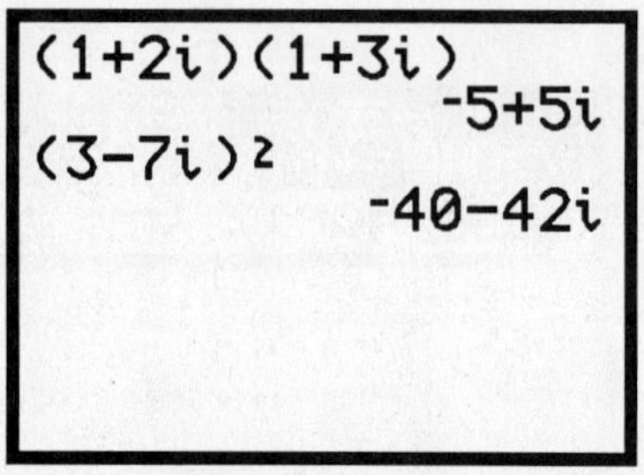

Complex numbers can also be divided in $a + bi$ mode. The real and imaginary parts of the quotient will be expressed in decimal form. For example, the quotient $(2 - 5i)/(1 - 6i)$ in Example 9, page 284 would be expressed as $.8648648649 + .1891891892i$.

Chapter 5
Analytic Geometry Topics

Many conic sections are represented by equations that are not functions. Consequently, these equations must be entered on the TI-82 and the TI-83 as two equations, each of which is a function.

GRAPHING PARABOLAS

To graph a parabola of the form $y^2 = 4px$ or $(y-k)^2 = 4p(x-h)$, we must first solve the equation for y.

Example 4, page 328: Graph the parabola $y^2 - 2y - 8x - 31 = 0$.

In the text we used the quadratic formula to solve the equation for y:

$$y = \frac{2 \pm \sqrt{32x+128}}{2}.$$

One way to produce the graph of the parabola is to enter $y_1 = \dfrac{2+\sqrt{32x+128}}{2}$ and $y_2 = \dfrac{2-\sqrt{32x+128}}{2}$, select a window, and press [GRAPH] to see the graph. Here we use $[-12, 12, -8, 8]$. The first equation produces the top half of the parabola and the second equation produces the lower half.

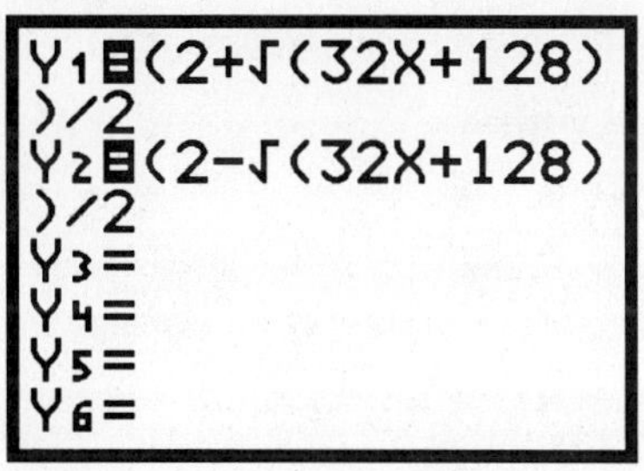

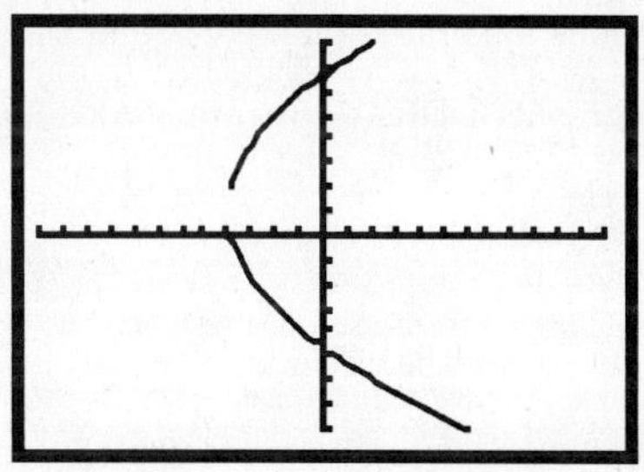

We can also enter $y_1 = \sqrt{32x+128}$. Then use the Y-VARS operation to enter $y_2 = \dfrac{2+y_1}{2}$ and $y_3 = \dfrac{2-y_1}{2}$. For example, to enter $y_2 = \dfrac{2+y_1}{2}$ on the TI-82 position the cursor beside "Y2 =" and press [(] 2 [+] [2nd] [Y-VARS] 1 1 [)] [÷] 2. On the TI-83 press [(] 2 [+] [VARS] [▷] 1 1 [)] [÷] 2. Enter $y_3 = \dfrac{2-y_1}{2}$ in a similar manner. Finally, deselect y_1 by moving the cursor to the equals sign following Y1 and pressing [ENTER]. The top half of the graph is produced by y_2 and the lower half by y_3. The expression for y_1 was entered to avoid entering the square root more than once. By deselecting y_1 we prevent its graph from appearing on the screen with the graph of the parabola.

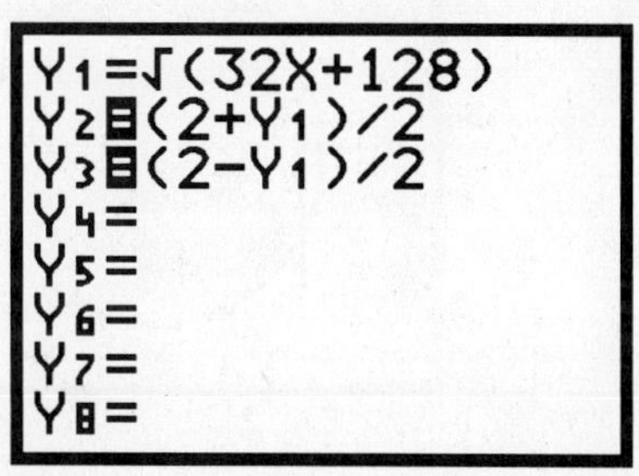

We could also use the standard equation of the parabola found in the text:

$$(y-1)^2 = 8(x+4).$$

Solve this equation for y.

$$y - 1 = \pm\sqrt{8(x+4)}$$
$$y = 1 \pm \sqrt{8(x+4)}$$

Then enter $y_1 = 1 + \sqrt{8(x+4)}$ and $y_2 = 1 - \sqrt{8(x+4)}$, or enter $y_1 = \sqrt{8(x+4)}$, $y_2 = 1 + y_1$, and $y_3 = 1 - y_1$, and deselect y_1 as described above.

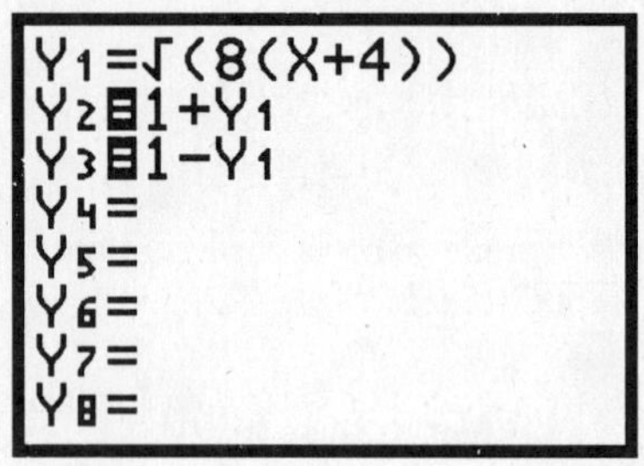

GRAPHING CIRCLES

One method for graphing a circle is illustrated on page 333 of the text. Another method can be used if the center and radius of a circle are known. The grapher's DRAW feature is used in this second method.

Example 1, page 332: For the circle

$$x^2 + y^2 - 16x + 14y + 32 = 0,$$

find the center and the radius. Then graph the circle..

In the text we completed the square twice to write the equation of the circle in standard form:

$$(x-8)^2 + [y-(-7)]^2 = 9^2.$$

Thus, the center of the circle is (8, −7),and the radius is 9. To graph it using the DRAW feature first press [Y =] and clear all previously entered equations. Then select a square window. We will use [−6, 24, −18, 2], Xscl = 2, Yscl = 2. Press [2nd] [QUIT] to go to the home screen. Then press [2nd] [DRAW] 9 to display "Circle(." Enter the coordinates of the center and the radius, separating the entries by commas, and close the parentheses: 8 [,] [(−)] 7 [,] 9 [)] [ENTER].

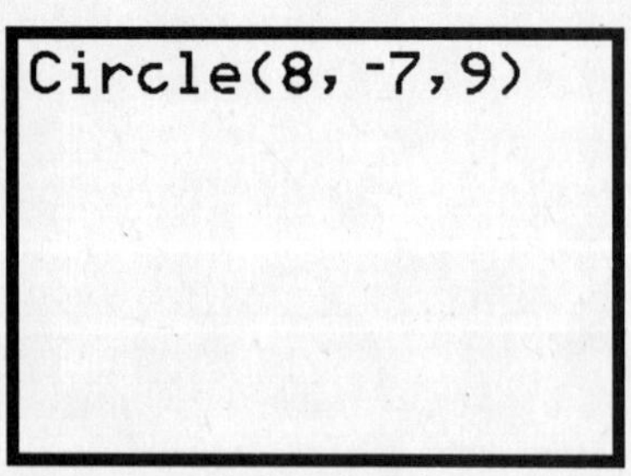

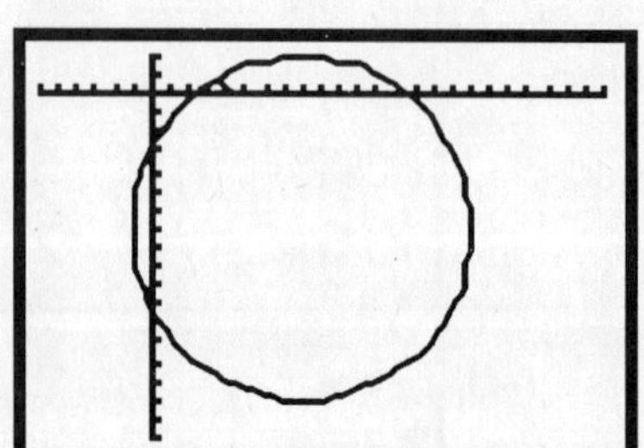

Note that the equation of a circle that is not centered at the origin, like the one in this example, is generally not easily solved for y. Since there is no need to solve for y using this method, it provides a quick procedure for graphing the equation of any circle in standard form. This drawing can be cleared as described on page 26 of this manual.

GRAPHING ELLIPSES

The equation of an ellipse must be solved for y before it can be entered on a TI-82 or a TI-83. In Example 2 of Section 5.2 of the text the procedure for graphing an ellipse of the form $\frac{x^2}{a^2}+\frac{y^2}{b^2}=1$ or $\frac{x^2}{b^2}+\frac{y^2}{a^2}=1$ is described. Here we consider ellipses of the form $\frac{(x-h)^2}{a^2}+\frac{(y-k)^2}{b^2}=1$ or $\frac{(x-h)^2}{b^2}+\frac{(y-k)^2}{a^2}=1$

Example 4, page 338: Graph the ellipse $4x^2+y^2+24x-2y+21=0$.

Completing the square in the text, we found that the equation can be written as

$$\frac{(x+3)^2}{4}+\frac{(y-1)^2}{16}=1.$$

Solve this equation for y.

$$\frac{(x+3)^2}{4}+\frac{(y-1)^2}{16}=1$$

$$\frac{(y-1)^2}{16}=1-\frac{(x+3)^2}{4}$$

$$(y-1)^2=16-4(x+3)^2 \qquad \text{Multiplying by 16}$$

$$y-1=\pm\sqrt{16-4(x+3)^2}$$

$$y=1\pm\sqrt{16-4(x+3)^2}$$

Now we can produce the graph in either of two ways. One is to enter $y_1 = 1+\sqrt{16-4(x+3)^2}$ and $y_2 = 1 - \sqrt{16-4(x+3)^2}$, select a square window, and press GRAPH. Here we use $[-9,9,-6,6]$. The first equation produces the top half of the ellipse and the second equation produces the lower half.

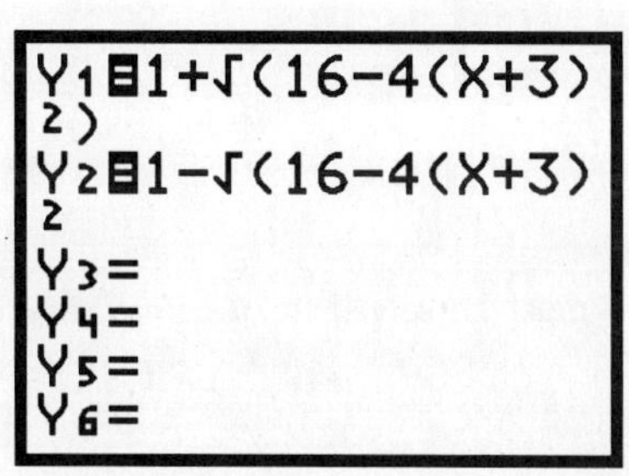

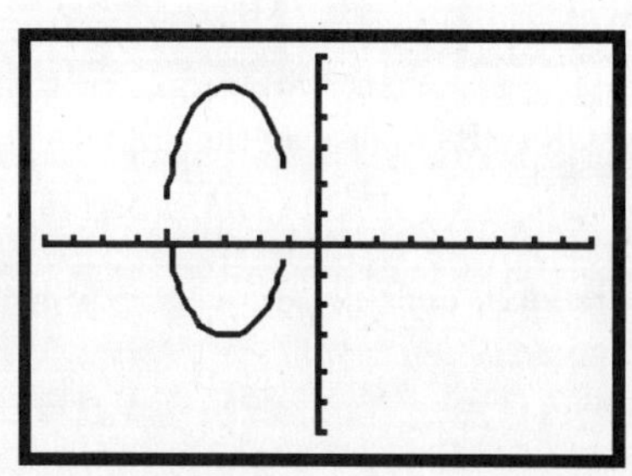

We can also enter $y_1 = \sqrt{16-4(x+3)^2}$ and use the Y-VARS operation as described above in Graphing Parabolas to enter $y_2 = 1+y_1$ and $y_3 = 1-y_1$. Then deselect y_1, select a square window, and press GRAPH. We use y_1 to eliminate the need to enter the square root more than once. Deselecting it prevents the graph of y_1 from appearing on the screen with the graph of the ellipse. The top half of the graph is produced by y_2 and the lower half by y_3.

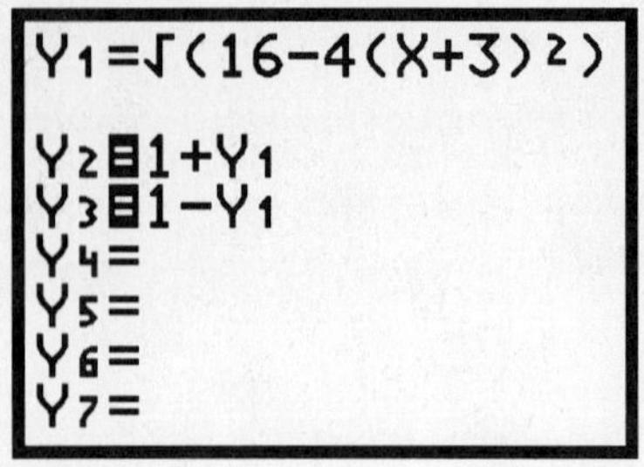

We could also begin by using the quadratic formula to solve the original equation for y.

$$\begin{aligned} 4x^2 + y^2 + 24x - 2y + 21 &= 0 \\ y^2 - 2y + (4x^2 + 24x + 21) &= 0 \\ y &= \frac{-(-2) \pm \sqrt{(-2)^2 - 4 \cdot 1 \cdot (4x^2 + 24x + 21)}}{2 \cdot 1} \\ y &= \frac{2 \pm \sqrt{4 - 16x^2 - 96x - 84}}{2} \\ y &= \frac{2 \pm \sqrt{-16x^2 - 96x - 80}}{2} \end{aligned}$$

Then enter $y_1 = \dfrac{2 + \sqrt{-16x^2 - 96x - 80}}{2}$ and $y_2 = \dfrac{2 - \sqrt{-16x^2 - 96x - 80}}{2}$, or enter $y_1 = \sqrt{-16x^2 - 96x - 80}$, $y_2 = \dfrac{2 + y_1}{2}$, and $y_3 = \dfrac{2 - y_1}{2}$, and deselect y_1.

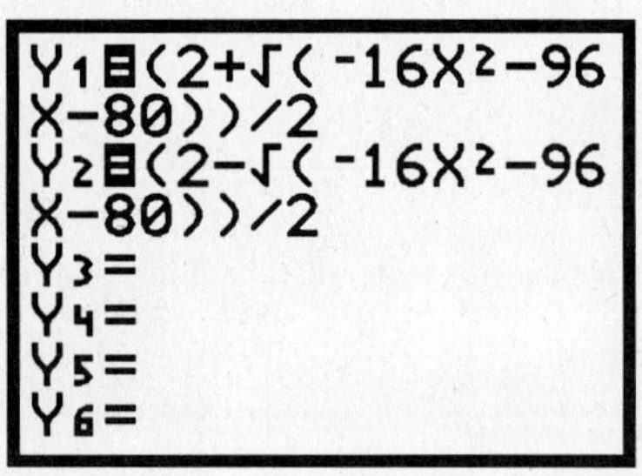

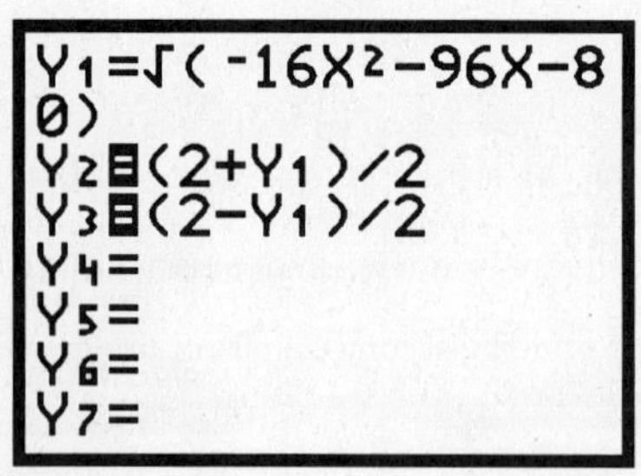

Select a square window and press GRAPH to display the graph.

GRAPHING HYPERBOLAS

As with equations of circles, parabolas, and ellipses, equations of hyperbolas must be solved for y before they can be entered on a TI-82 or a TI-83.

Example 2, page 344: Graph the hyperbola $9x^2 - 16y^2 = 144$.

First solve the equation for y.

$$\begin{aligned} 9x^2 - 16y^2 &= 144 \\ -16y^2 &= -9x^2 + 144 \\ y^2 &= \frac{-9x^2 + 144}{-16} \\ y &= \pm\sqrt{\frac{-9x^2 + 144}{-16}}, \text{ or } \pm\sqrt{\frac{9x^2 - 144}{16}} \end{aligned}$$

It is not necessary to simplify further.

Now enter $y_1 = \sqrt{\dfrac{9x^2 - 144}{16}}$ and either $y_2 = -\sqrt{\dfrac{9x^2 - 144}{16}}$ or $y_2 = -y_1$, select a square window, and press GRAPH.

Here we use $[-9, 9, -6, 6]$. The top half of the graph is produced by y_1 and the lower half by y_2.

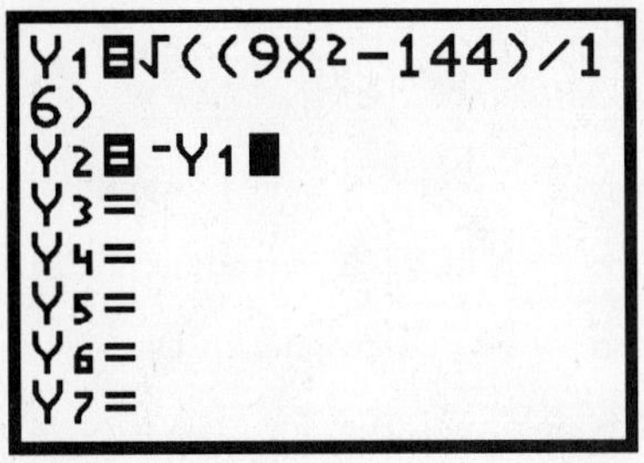

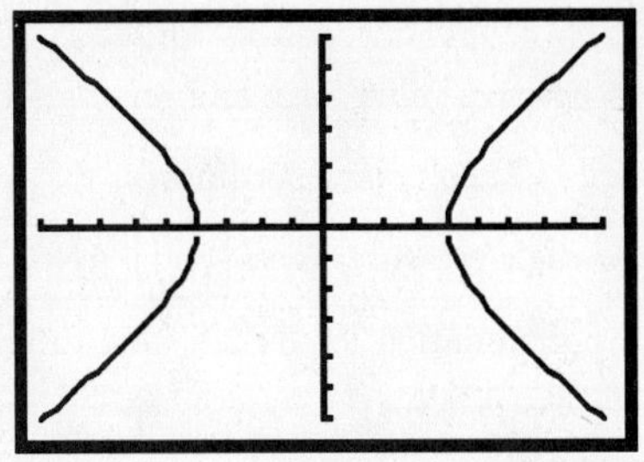

Example 3, page 347: Graph the hyperbola $4y^2 - x^2 + 24y + 4x + 28 = 0$.

In the text we completed the square to get the standard form of the equation. Now solve the equation for y.

$$\frac{(y+3)^2}{1} - \frac{(x-2)^2}{4} = 1$$

$$(y+3)^2 = \frac{(x-2)^2}{4} + 1$$

$$y + 3 = \pm\sqrt{\frac{(x-2)^2}{4} + 1}$$

$$y = -3 \pm \sqrt{\frac{(x-2)^2}{4} + 1}$$

The graph can be produced in either of two ways. One is to enter $y_1 = -3 + \sqrt{\frac{(x-2)^2}{4} + 1}$ and $y_2 = -3 - \sqrt{\frac{(x-2)^2}{4} + 1}$, select a square window, and press GRAPH. Here we use $[-12, 12, -9, 9]$. The first equation produces the top half of the hyperbola and the second the lower half.

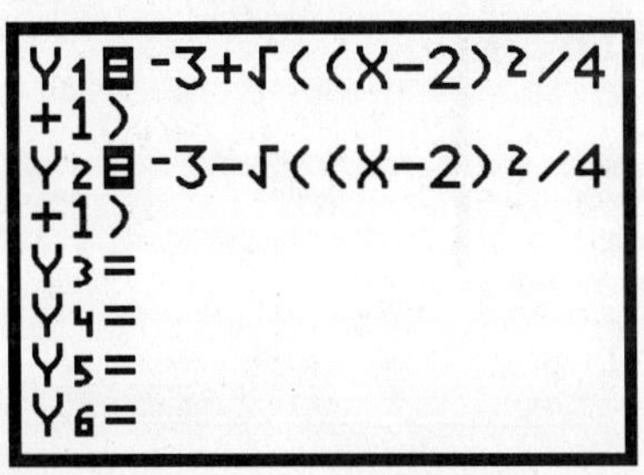

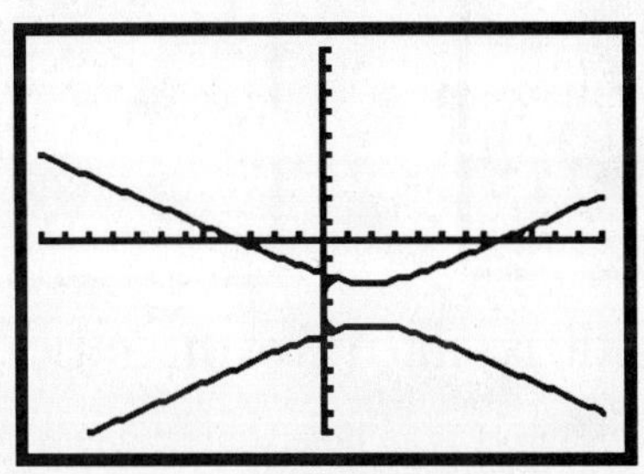

We can also enter $y_1 = \sqrt{\frac{(x-2)^2}{4} + 1}$, $y_2 = -3 + y_1$, and $y_3 = -3 - y_1$. Then deselect y_1, select a square window, and press GRAPH. Again, y_1 is used to eliminate the need to enter the square root more than once. Deselecting it prevents the graph of y_1 from appearing on the screen with the graph of the hyperbola. The top half of the graph is produced by y_2 and the lower half by y_3.

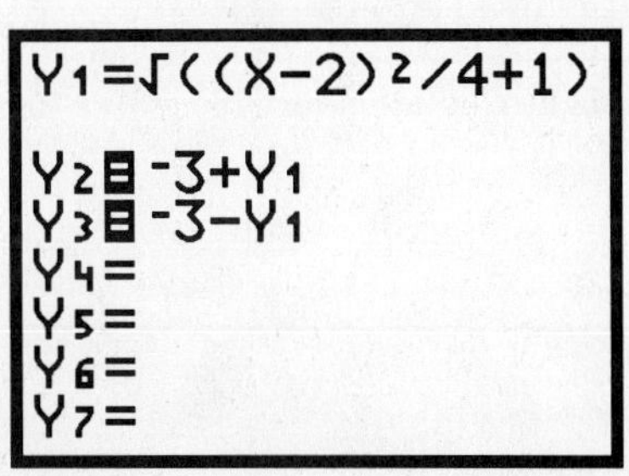

CONVERTING FROM RECTANGULAR TO POLAR COORDINATES

The grapher can be used to convert from rectangular to polar coordinates, expressing the result using either degrees or radians. The grapher will supply a positive value for r and an angle in the interval $(-180°, 180°]$, or $(-\pi, \pi]$.

Example 2 (a), page 370: Convert (3,3) to polar coordinates.

To find r, regardless of the type of angle measure, press [2nd] [ANGLE] 5 3 [,] 3 [)] [ENTER]. The readout is 4.242640687, so $r \approx 4.2426$. This is a decimal approximation for $3\sqrt{2}$. Now, to find θ in degrees, set the grapher in Degree mode and press [2nd] [ANGLE] 6 3 [,] 3 [)] [ENTER]. The readout is 45, so $\theta = 45°$. Thus polar notation for (3,3) is (4.2426, 45°).

Set the grapher in Radian mode to find θ in radians. Repeat the keystrokes for finding θ above to find that $\theta \approx 0.7854$. This is a decimal approximation for $\pi/4$. Thus polar notation for (3,3) is (4.2426,0.7854).

CONVERTING FROM POLAR TO RECTANGULAR COORDINATES

The grapher can also be used to convert from polar to rectangular coordinates.

Example 3, page 371: Convert each of the following to rectangular coordinates.

(a) $(10, \pi/3)$ (b) $(-5, 135°)$

(a) Since the angle is given in radians, set the grapher in Radian mode. To find the x-coordinate of rectangular notation, press [2nd] [ANGLE] 7 1 0 [,] [2nd] [π] [÷] 3 [)] [ENTER]. The readout is 5, so $x = 5$. The y-coordinate is found by pressing [2nd] [ANGLE] 8 1 0 [,] [2nd] [π] [÷] 3 [)] [ENTER]. The readout is 8.660254038, so $y \approx 8.6603$. This is a decimal approximation of $5\sqrt{3}$. Thus, rectangular notation for $(10, \pi/3)$ is (5,8.6603).

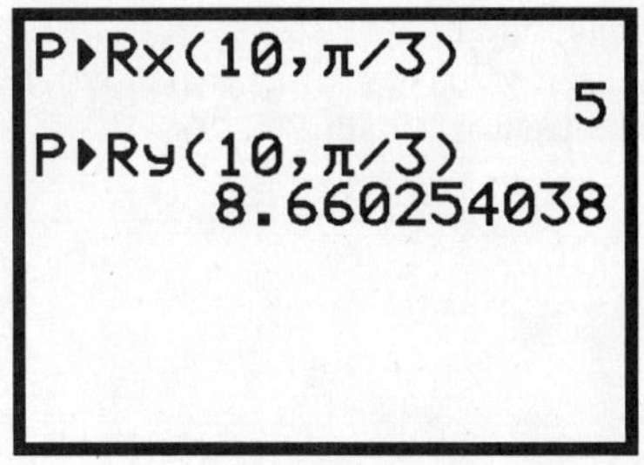

(b) The angle is given in degrees, so we set the grapher in Degree mode. To find the x-coordinate of rectangular notation, press [2nd] [ANGLE] 7 [(−)] 5 [,] 1 3 5 [)] [ENTER]. The readout is 3.535533906, so $x \approx 3.5355$. This is a decimal approximation of $\frac{5\sqrt{2}}{2}$. The y-coordinate is found by pressing [2nd] [ANGLE] 8 [(−)] 5 [,] 1 3 5 [)] [ENTER]. The readout is −3.535533906, so $y \approx -3.5355$. This is a decimal approximation of $-\frac{5\sqrt{2}}{2}$. Thus, rectangular notation for (−5,135°) is (3.5355, −3.5355).

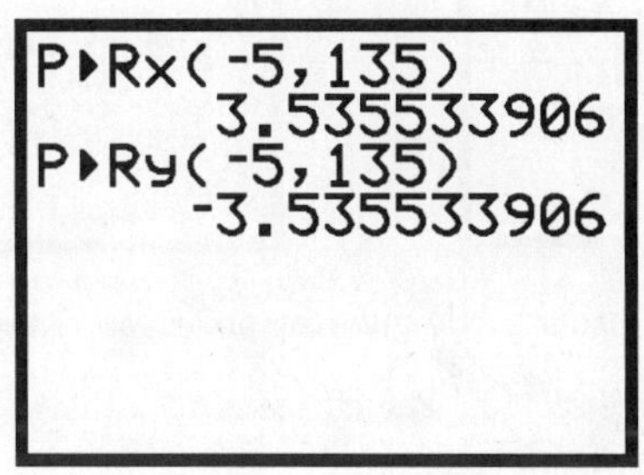

GRAPHING POLAR EQUATIONS

Polar equations can be graphed in either Radian mode or Degree mode. The equation must be written in the form $r = f(\theta)$ and the grapher must be set in Polar (Pol) mode. Typically we begin with a range of $[0, 2\pi]$ or $[0°, 360°]$, but it might be necessary to increase the range to ensure that sufficient points are plotted to display the entire graph.

Example 6, page 373: Graph: $r = 1 - \sin\theta$.

First set the grapher in Polar mode by pressing [MODE] [▽] [▽] [▽] [▷] [▷] [ENTER]. We will also select Radian mode.

The equation is given in $r = f(\theta)$ form. Press [Y =] to enter it on the "Y =" screen. Clear any existing entries and, with the cursor beside "r_1 =," press 1 [−] [SIN] [X,T,θ]. Now press [WINDOW] and enter the following settings:

θmin = 0 (Smallest value of θ to be evaluated)
θmax = 2π (Largest value of θ to be evaluated)
θstep = $\pi/24$ (Increment in θ values)
Xmin = −3
Xmax = 3
Xscl = 1
Ymin = −3
Ymax = 1
Yscl = 1

With these settings the grapher evaluates the function from $\theta = 0$ to $\theta = 2\pi$ in increments of $\pi/24$ and displays the graph in the square window $[-3, 3, -3, 1]$. Values entered in terms of π appear on the screen as decimal approximations. Press GRAPH to display the graph.

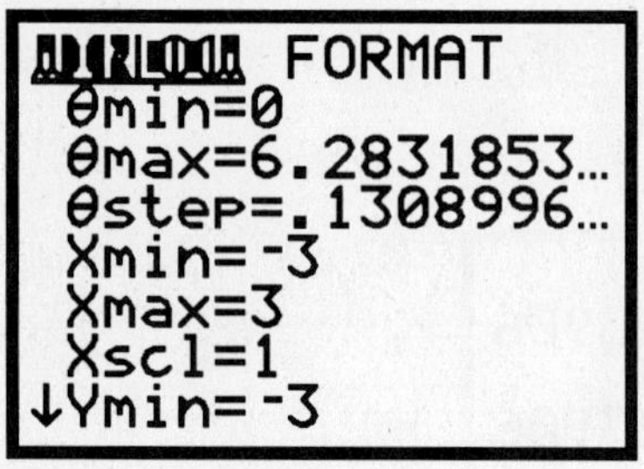

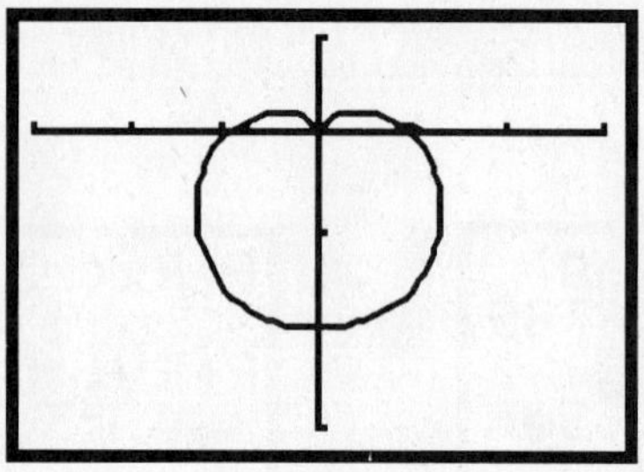

The curve can be traced with either rectangular or polar coordinates being displayed. The value of θ is also displayed when rectangular coordinates are selected. The choice of coordinates is made on the WINDOW FORMAT screen. Press WINDOW ▷ on the TI-82 or 2nd FORMAT on the TI-83 to display this screen. (FORMAT is the second operation associated with the ZOOM key.) Then position the blinking cursor over RectGC to select rectangular coordinates or over PolarGC to select polar coordinates and press ENTER.

Chapter 6
Exponential and Logarithmic Functions

EVALUATING e^x, Log x, and Ln x

Use the grapher's scientific keys to evaluate e^x, $\log x$, and $\ln x$ for specific values of x.

Example 6 (a), (b), page 407: Find the value of e^3 and $e^{-0.23}$. Round to four decimal places.

To find e^3 press [2nd] [e^x] 3 [ENTER]. (e^x is the second operation associated with the [LN] key.) The grapher returns 20.08553692. Thus, $e^3 \approx 20.0855$. To find $e^{-0.23}$ press [2nd] [e^x] [(−)] [.] 2 3 [ENTER]. The grapher returns .7945336025, so $e^{-0.23} \approx 0.7945$.

Example 4 (a), (b), (c), page 416: Find the values of log 645,778, log 0.0000239, and log (−3). Round to four decimal places.

To find log 645,778 press [LOG] 6 4 5 7 7 8 [ENTER] and read 5.810083246. Thus, $\log 645,778 \approx 5.8101$. To find log 0.0000239 press [LOG] [.] 0 0 0 0 2 3 9 [ENTER]. The grapher returns −4.621602099, so $\log 0.0000239 \approx -4.6216$. When we press [LOG] [(−)] 3 [ENTER] on the TI-82 the message ERR: DOMAIN is returned. This indicates that −3 is not in the domain of the function $\log x$. When the TI-83 is set in Real mode the keystrokes [LOG] [(−)] 3 [ENTER] produce the message ERR: NONREAL ANS indicating that the result of this calculation is not a real number.

Example 5 (a), (b), (c), page 416: Find the values of ln 645,778, ln 0.0000239, and ln (−5). Round to four decimal places.

To find ln 645,778 and ln 0.0000239 repeat the keystrokes used above to find log 645,778 and log 0.0000239 but press [LN] rather than [LOG]. We find that ln 645,778 ≈ 13.3782 and $\ln 0.0000239 \approx -10.6416$. When we press [LN] [(−)] 5 [ENTER] on the TI-82 the message ERR: DOMAIN is returned indicating that −5 is not in the domain of the function $\ln x$. When the TI-83 is set in Real mode these keystrokes produce the message ERR: NONREAL ANS indicating that the result of this calculation is not a real number.

USING THE CHANGE OF BASE FORMULA

To find a logarithm with a base other than 10 or e we use the change-of-base formula, $\log_b M = \dfrac{\log_a M}{\log_a b}$, where a and b are any logarithmic bases and M is any positive number.

Example 6, page 417: Find $\log_5 8$ using common logarithms.

We let $a = 10$, $b = 5$, and $M = 8$ and substitute in the change-of-base formula. Press [LOG] 8 [÷] [LOG] 5 [ENTER]. On the TI-83 it is necessary to close the parentheses in the numerator to enter the expression correctly. Press [LOG] 8 [)] [÷] [LOG] 5 [ENTER]. In either case the result is about 1.2920.

Example 7, page 417: Find $\log_4 31$ using natural logarithms.

We let $a = e$, $b = 4$, and $M = 31$ and substitute in the change-of-base formula. On the TI-82 press [LN] 3 1 [÷] [LN] 4

ENTER. A right parenthesis must be added after the 1 on the TI-83. The result is about 2.4771.

Example 9, page 418: Graph $y = \log_5 x$.

To use a grapher we must first change the base to e or 10. Here we use e. Let $a = e$, $b = 5$, and $M = x$ and substitute in the change-of-base formula. Enter $y_1 = \dfrac{\ln x}{\ln 5}$ on the "Y =" screen, select a window, and press GRAPH. Note that since the TI-83 forces the use of parentheses with the ln function, the parentheses in the numerator must be closed on this grapher: $\ln(x)/\ln(5)$. The right parenthesis following the 5 is optional.

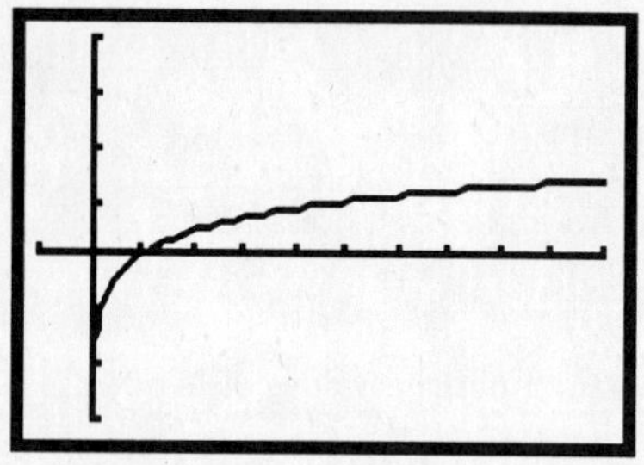

EXPONENTIAL, LOGARITHMIC, AND POWER REGRESSION

Exponential, logarithmic, and power functions can be fit to data. The operations of entering data, making scatterplots, and graphing and evaluating these functions are the same as for linear regression functions. So are the procedures for copying a regression equation to the "Y =" screen, graphing it, and using it to find function values. See Appendix A.3 of this manual (page 53) for a discussion of these procedures. Note that the coefficient of correlation, r, will be displayed on the TI-83 only if DiagnosticOn has been selected from the CATALOG.

Example 6 (a), page 447: Fit an exponential equation to the given data on cellular phones.

Enter the data in L_1 and L_2 as described on page 53 of this manual. Then press STAT ▷ to view the STAT CALC menu. Select ExpReg by pressing ALPHA A on the TI-82 or 0 on the TI-83 followed by ENTER. The values of a and b for the exponential function $y = ab^x$ are displayed along with the coefficient of correlation r. The TI-83 also displays r^2.

Exercise 26 (a), page 453: Fit a logarithmic function to the given data on forgetting.

After entering the data in L_1 and L_2 as described on page 53 of this manual, press STAT ▷ to view the STAT CALC menu. Select LnReg by pressing 0 on the TI-82 or 9 on the TI-83 followed by ENTER. The values of a and b for the logarithmic function $y = a + b \ln x$ are displayed along with the coefficient of correlation r. The TI-83 also displays r^2.

Example 7 (a), page 449: Fit a power function to the given data on cholesterol level and risk of heart attack.

Enter the data in L_1 and L_2 as described on page 53 of this manual. Then press STAT ▷ to view the STAT CALC menu. Select PwrReg by pressing ALPHA B on the TI-82 or ALPHA A on the TI-83 followed by ENTER. The values of a and b for the power function $y = ax^b$ are displayed along with the coefficient of correlation r. The TI-83 also displays r^2.

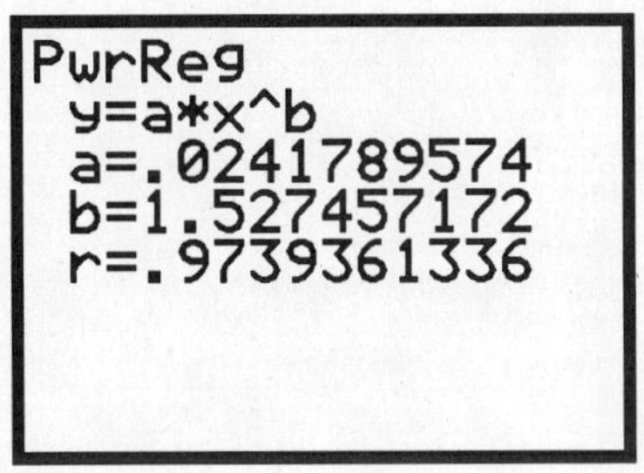

LOGISTIC REGRESSION

A logistic function can be fit to data using the TI-83.

Exercise 28 (a), page 452: Fit a logistic function to the given data on the effect of advertising.

After entering the data in L_1 and L_2 as described on page 53 of this manual, press STAT ▷ to view the STAT CALC menu. Select Logistic by pressing ALPHA B ENTER. The values of a, b, and c for the logistic function $y = \dfrac{c}{1 + ae^{-bx}}$ are displayed.

Appendix A.1
Solving Equations

CHECKING SOLUTIONS USING A TABLE

The grapher's TABLE feature can be used to check solutions of equations. In this situation it is most efficient to select "Ask" for the Indpnt setting in the TABLE SETUP screen. To do this, press 2nd TblSet ▽ ▽ ▷ ENTER. (The values of TblMin and ΔTbl are irrelevant in Ask mode.)

Now press 2nd TABLE. The grapher displays a table that allows us to enter values for the independent variable. We can then enter the left-hand side of an equation as y_1 and the right-hand side as y_2 in the "Y =" screen and use the TABLE feature to evaluate each side for any values that are possible solutions of the equation. For a given value of x, when $y_1 = y_2$ the equation is true and the given x-value is a solution . When $y_1 \neq y_2$ the equation is false and the given x-value is not a solution.

Example 1, page 460: The possible solution of $3(7 - 2x) = 14 - 8(x - 1)$ is $1/2$. Use a table to check.

Enter $y_1 = 3(7 - 2x)$ and $y_2 = 14 - 8(x - 1)$ in the "Y =" screen. Now, with the table set in Ask mode, press 2nd TABLE to display the empty table and then press 1 ÷ 2 ENTER or .5 ENTER. We see that Y1 = Y2 = 18, so the equation is true when X = 1/2 and thus the solution is 1/2.

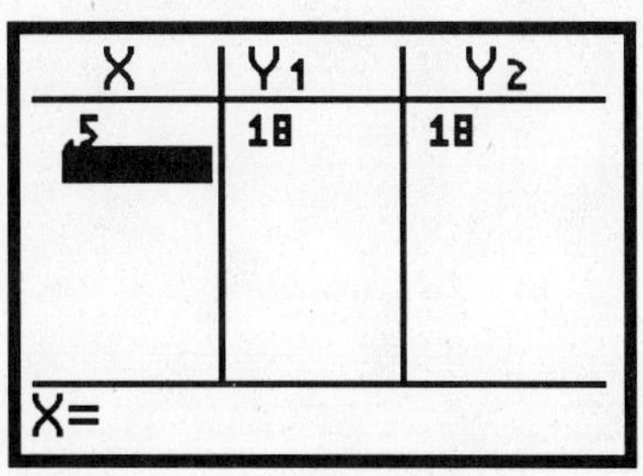

Example 7, page 466: The possible solutions of $5 + \sqrt{x+7} = x$ are 9 and 2. Use a table to check.

Enter $y_1 = 5 + \sqrt{x+7}$ and $y_2 = x$. With the table set in Ask mode, display the TABLE screen and press 9 ENTER 2 ENTER. The table shows that Y1 = Y2 = 9 when X = 9, but Y1 = 8 and Y2 = 2 when X = 2. Thus, 9 is a solution of the equation, but 2 is not.

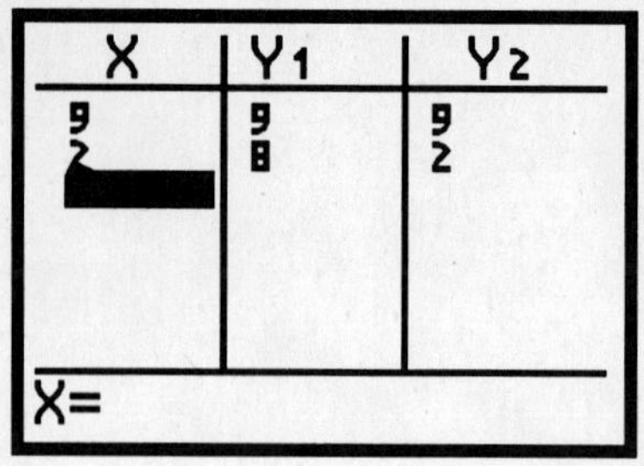
X
Y1
Y2
X=

Appendix A.2
Distance, Midpoints, and Circles

GRAPHING CIRCLES

Although the standard form of the equation of a circle cannot be entered in the "Y =" screen of the TI-82 or the TI-83, circles can be graphed on these graphers. A square viewing window is required to give an accurate graph using either of the methods discussed below.

Method 1: We begin by solving the equation for y.

Example 5, page 474: Graph $x^2 + y^2 = 16$.

Solve the equation for y as shown on page 474 of the text, enter $y_1 = \sqrt{16 - x^2}$ and $y_2 = -\sqrt{16 - x^2}$, select a square window, and press [GRAPH]. Note that we can enter $y_2 = -y_1$ instead of $y_2 = -\sqrt{16 - x^2}$. The graph is shown on page 474 of the text.

Method 2: If the center and radius of a circle are known, the circle can be graphed using the DRAW feature.

Exercise 37, page 475: Graph $(x - 1)^2 + (y - 5)^2 = 36$.

The center of this circle is (1,5) and its radius is 6. To graph it using the DRAW feature first press [Y =] and clear all previously entered equations. Then select a square window. We will use $[-12, 12, -4, 12]$, Xscl = 2, Yscl = 2. Press [2nd] [QUIT] to go to the home screen. Then press [2nd] [DRAW] 9 to display "Circle(." Enter the coordinates of the center and the radius, separating the entries by commas, and close the parentheses: 1 [,] 5 [,] 6 [)] [ENTER].

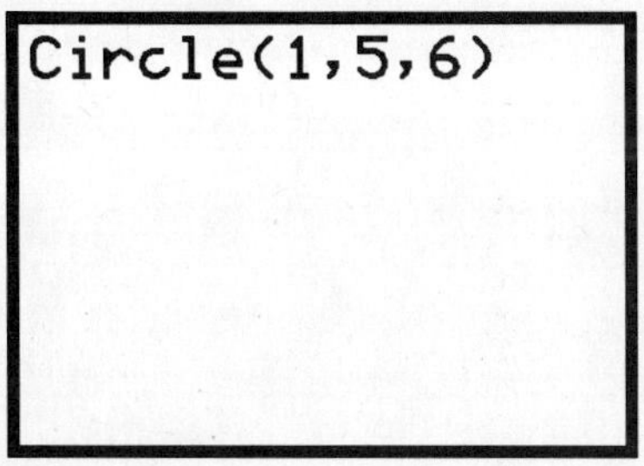

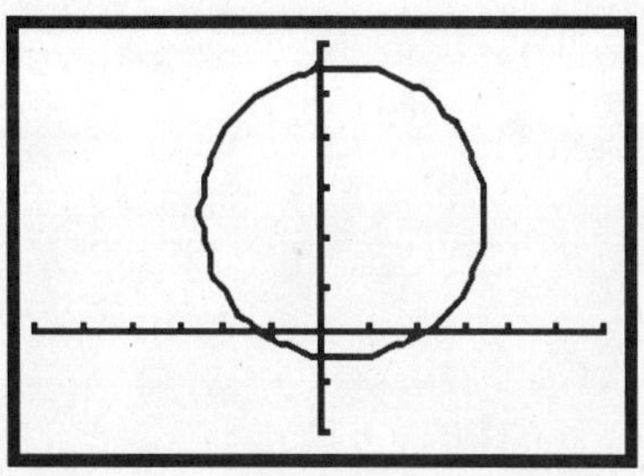

Note that the equation of a circle that is not centered at the origin, like the one in this example, is generally not easily solved for y, making Method 2 preferable for such equations. In fact, since there is no need to solve for y using this method, it provides a quick procedure for graphing the equation of any circle in standard form. This drawing can be cleared as described on page 26 of this manual.

Appendix A.3
Data Analysis, Curve Fitting, and Regression

LINEAR REGRESSION AND SCATTERPLOTS

The linear regression operation of the grapher enables us to fit a linear function to a set of data.

Example 1, page 478: Fit a regression line to the data on life expectancy of women on page 477 of the text. Then predict the life expectancy of women in 2000.

First we enter the data from the table on the STAT list editor screen. To clear any existing lists press [STAT] 4 [2nd] [L_1] [,] [2nd] [L_2] [,] [2nd] [L_3] [,] [2nd] [L_4] [,] [2nd] [L_5] [,] [2nd] [L_6] [ENTER]. (L_1 through L_6 are the second operations associated with the numeric keys 1 through 6.) Now press [STAT] 1 to display the STAT list editor. We will enter the x-values in L_1 and the y-values in L_2. The cursor should be at the top of the L_1 column. To enter 0 press 0 [ENTER]. Continue typing the x-values 1 through 9, each followed by [ENTER]. Press [▷] to move to the top of the L_2 column. Type the y-values in succession, each followed by [ENTER].

To see a scatterplot of the data first press [2nd] [STAT PLOT] 1. (STAT PLOT is the second operation associated with the [Y =] key.) Position the cursor over On and press [ENTER] to turn on the scatterplot of these data. The entries Type, Xlist, and Ylist should be as shown below on the left for the TI-82 and on the right for the TI-83. The last item, Mark, allows us to choose a box, a cross, or a dot for each point of the scatterplot. Here we have selected a box. All of these entries except for Xlist and Ylist on the TI-83 are chosen by positioning the cursor over each selection and pressing [ENTER]. Use the L_1 through L_6 keys to define Xlist and Ylist on the TI-83.

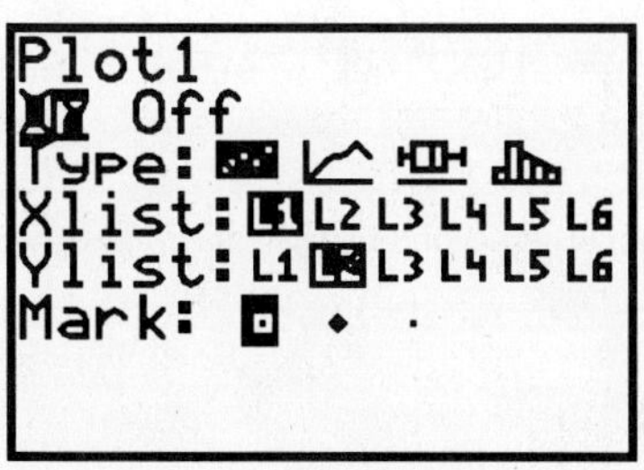

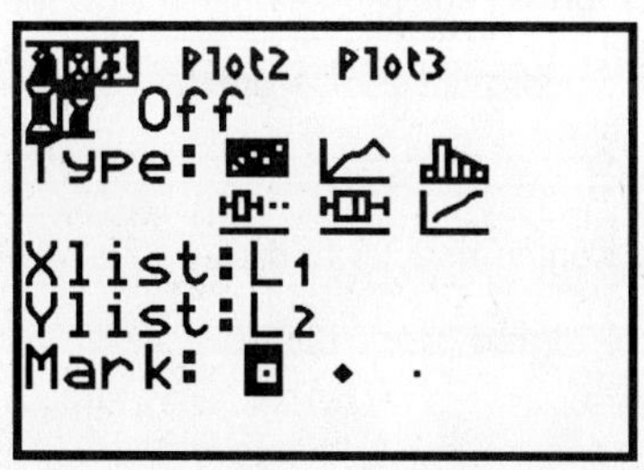

On the TI-83 the scatterplots can also be turned on from the "Y =" screen. In this case we would position the cursor over Plot 1 and press [ENTER]. Plot 1 will now be highlighted.

To see the scatterplot select a window and press [GRAPH]. Instead of entering the window dimensions directly, we can press [ZOOM] 9. This activates the ZoomStat operation which automatically defines a viewing window that displays all the data points. The STAT lists and the scatterplot for these data are shown on page 478 of the text.

Press [STAT] [▷] 5 [ENTER] on the TI-82 or [STAT] [▷] 4 [ENTER] on the TI-83 to select LinReg(ax+b) from the STAT CALC menu and to display the coefficients a and b of the regression line $y = ax + b$ and the coefficient of linear correlation, r. Note that DiagnosticOn mode must have been selected from the CATALOG on the TI-83 in order for r to be displayed. In DiagnosticOn mode the TI-83 will, in fact, display both r^2 and r.

To select this mode, press 2nd CATALOG and use ▽ to position the triangular selection cursor beside DiagnosticOn. To alleviate the tedium of scrolling through many items to reach DiagnosticOn, press D after pressing 2nd CATALOG to move quickly to the first catalog item that begins with the letter D. (D is the ALPHA operation associated with the x^{-1} key.) Then use ▽ to scroll to DiagnosticOn. Note that it is not necessary to press ALPHA before D when the catalog is displayed. Press ENTER to paste this instruction to the home screen and then press ENTER a second time to set the mode.

```
LinReg
 y=ax+b
 a=3.427878788
 b=50.65454545
 r=.9861247538
```

```
LinReg
 y=ax+b
 a=3.427878788
 b=50.65454545
 r²=.9724420301
 r=.9861247538
```

To enter the regression equation as y_1 in the "Y =" screen press Y = and position the cursor beside "Y1 =". Then on the TI-82 press VARS 5 ▷ ▷ 7 and on the TI-83 press VARS 5 ▷ ▷ 1. The TI-83 also offers the option of specifying a y-variable to which the regression equation is stored before the equation is found. Store the equation as y_1, for example, by pressing VARS ▷ 1 1 after STAT ▷ 4 is pressed. The screen will display LinReg Y1. Then press ENTER to see the regression equation and to store it as y_1.

To graph the regression line on the same screen as the scatterplot press GRAPH. (See page 478 of the text for this graph.) To turn off the scatterplot press 2nd STAT PLOT 1, position the cursor over Off, and press ENTER. On the TI-83 we can also position the cursor over Plot 1 on the "Y =" screen and press ENTER. Now Plot 1 is no longer highlighted.

To predict the life expectancy of women in 2000 find the value of y_1 when $x = 10$ using one of the methods described in "Finding Function Values" on pages 23 and 24 of this manual.

QUADRATIC, CUBIC, AND QUARTIC REGRESSION

Second degree, third degree, and fourth degree polynomial functions can be fit to data using the quadratic, cubic, and quartic regression operations, respectively, on the TI-82 and the TI-83.

The operations of entering data, making scatterplots, and graphing and evaluating quadratic, cubic, and quartic regression functions are the same as for linear regression functions.

Example 2, page 481: For the data on hours of sleep versus death rate on page 481 of the text, make a scatterplot for the data and determine which, if any, of the functions above fits the data. Then find the regression equation, graph the equation, and use it to make predictions.

Enter the data in L_1 and L_2 and make a scatterplot as described above for linear regression. The scatterplot is shown in the text. It appears that a quadratic function fits the data. To find it press STAT ▷ to view the STAT CALC menu. Then select QuadReg by pressing 6 on the TI-82 or 5 on the TI-83 followed by ENTER. The coefficients of the quadratic equation $y = ax^2 + bx + c$ are displayed. Note that at least three data points are required for quadratic regression.

The regression equation can be copied to the "Y =" screen using the VARS operation as described on page 54 of this manual. The TI-83 also offers the usual option of selecting a y-variable name before the regression equation is found. Once the regression equation is entered in the "Y =" screen, it can be graphed and used to make predictions using one of the methods for finding function values described on pages 23 and 24 of this manual.

To fit a cubic or quartic function to a set of data, enter the data in L_1 and L_2 as described earlier. Then use STAT ▷ to go to the STAT CALC menu. On the TI-82 we select cubic regression by pressing 7 and quartic regression by pressing 8. Press 6 for cubic regression on the TI-83 and 7 for quartic regression. At least four data points are required for cubic regression and at least five are required for quartic regression.

Programs For The TI-82 Graphics Calculator

PROGRAMS FOR THE TI-82 GRAPHICS CALCULATOR

(Created by Dave Slomer, Winton Woods High School, Cincinnati, OH.)

QUADFORM

This program solves a quadratic equation of form $Ax^2 + Bx + C = 0$.

INPUT: A, B, C

OUTPUT: Real or complex solutions

```
:Prompt A,B,C
:B^2-4*A*C→D
:Disp "ROOTS:"
:If D≥Ø:Then
:Disp (-B+√(D))/(2*A)
:Disp (-B-√(D))/(2*A)
:Else
:Disp -B/(2*A),"+/-",√(-D)/(2*A),"I"
:End
```

TRISOL9Ø

This program solves right triangles ABC in which legs L and M are opposite angles A and B, respectively.

INPUT: One side and any other part (except the right angle, C)

OUTPUT: First, the sides; then, the acute angles

```
:Degree
:Menu("TRISOL9Ø","HA",P,"HL",Q,"LA",R,"LL",S)
:Lbl P
:Prompt H,A
:9Ø-A→B
:H*sin (A)→L
:H*sin (B)→M
:Goto X
:Lbl Q
:Prompt H,L
:√(H^2-L^2)→M
:sin⁻¹ (L/H)→A
:9Ø-A→B
:Goto X
:Lbl R
:Prompt L,A
:9Ø-A→B
:L/tan (A)→M
:√(L^2+M^2)→H
:Goto X
:Lbl S
:Prompt L,M
:√(L^2+M^2)→H
:tan⁻¹ (L/M)→A
:9Ø-A→B
:Lbl X
:Disp L,M,H,A,B
```

TRISOL

This program solves triangle ABC, in which sides L, M, and N are opposite angles A, B, and C, respectively. It covers the "no solution" case as well as the "two solutions" case.

INPUT: Three parts of a triangle, including at least one side

OUTPUT: First the sides, then the angles. The AAS case is not included, so in this case it is necessary to find the third angle by hand and then use ASA.

```
:Degree
:Disp "ANGLES A,B,C"
:Disp "SIDES L,M,N"
:Menu("TRISOL","SSS",P,"SAS",Q,"(A)ASA",R,"SSA",S)
:Lbl P
:Prompt L,M,N
:cos⁻¹ ((M^2+N^2-L^2)/(2*M*N))→A
:cos⁻¹ ((L^2+N^2-M^2)/(2*L*N))→B
:180-A-B→C
:Goto X
:Lbl Q
:Prompt L,C,M
:√(L^2+M^2-2*L*M*cos (C))→N
:cos⁻¹ ((M^2+N^2-L^2)/(2*M*N))→A
:180-A-C→B
:Goto X
:Lbl R
:Prompt A,N,B
:180-A-B→C
:sin (A)*N/(sin (C))→L
:sin (B)*N/(sin (C))→M
:Goto X
:Lbl S
:Prompt L,M,B
:L*sin (B)/M→Z
:If Z>1:Then:Disp "NO SOL":Stop:End
:sin⁻¹ (Z)→A
:180-A-B→C
:sin (C)*M/(sin (B))→N
:Disp L,M,N,A,B,C
:If B≥A:Stop
:Disp "2ND SOL...":Pause
:180-A→A
:180-A-B→C
:sin (C)*M/(sin (B))→N
:Lbl X
:Disp L,M,N,A,B,C
```

DEMOIVRE

This program helps visualize Nth roots of a complex number $A + Bi$.

INPUT: A, B, N

OUPUT: Traceable graph of the N points (A, B) corresponding to the roots

```
:Param
:Degree
:Input "REAL PART:",A
:Input "IMAG PART:",B
:Input "INDEX:",N
:√ (A²+B²)^(1/N)→R
:If A=Ø:Then
:18Ø-abs (B)/B*9Ø
:Else
:tan⁻¹ (B/A)+18Ø*(A<Ø)
:End
:Ans/N→H
:"R*cos (T)"→X1T
:"R*sin (T)"→Y1T
:H→Tmin
:H+36Ø→Tmax
:36Ø/N→Tstep
:1.5R→Ymax
:-Ymax→Ymin
:1.5Ymax→Xmax
:-Xmax→Xmin
:Trace
```

LOGBASE

This program computes logarithms of any base.

INPUT: Number (X) and base (B)

OUTPUT: $\log_B(X)$

```
:Prompt X
:Input "NEW BASE:",B
:log (X)/log (B)
```

LOAN

This program computes various parts of mortgage payments via the "present value" formula. $L = PM(1-(1+R/M)^{-MY})/R$, whose parameters are defined as follows:

L = the loan amount (wanted now, in the present),
P = the periodic loan payment,
R = the nominal (stated) interest rate,
M = the number of annual conversions, and
Y = the number of years of the loan.

INPUT: Menu choices about which parameter to change and which to find
OUTPUT: Values of all parameters

```
:"L-P*M*(1-(1+R/M)^-(M*Y))/R"→YØ
:Func:FnOff Ø
:Lbl Ø
:ClrHome
:Disp "L,P,R,Y,M:"
:Disp L,P,R,Y,M:Pause
:Menu("CHANGE:","(L)OAN AMT",1,"(P)AYMENT AMT",2,"(R)ATE OF INT",3,"(Y)EARS",4,"ANN.
CONV. (M)",5,"<<COMPUTE>>",7,"<<REVIEW>>",Ø)
:Lbl 1:Prompt L:Goto Ø
:Lbl 2:Prompt P:Goto Ø
:Lbl 3:Prompt R:Goto Ø
:Lbl 4:Prompt Y:Goto Ø
:Lbl 5:Prompt M:Goto Ø
:Lbl 7
:Menu("FIND:","LOAN AMT",A,"PAYMENT AMT",B,"RATE OF INT",C,"YEARS",D,"<<RETURN>>",Ø)
:Lbl A:solve(YØ,L,1)→L:Goto 8
:Lbl B:solve(YØ,P,1)→P:Goto 8
:Lbl C:solve(YØ,R,1)→R:Goto 8
:Lbl D:solve(YØ,Y,1)→Y
:Lbl 8:Pause Ans:Goto Ø
```

The TI-85 Graphics Calculator

Preliminaries

Press [ON] to turn on the TI-85 graphing calculator. ([ON] is the key at the bottom left-hand corner of the keypad.) The display contrast can be adjusted by first pressing [2nd]. ([2nd] is the yellow key in the left column of the keypad.) Then press and hold [△] to increase the contrast or [▽] to decrease the contrast. To turn the grapher off, press [2nd] [OFF]. (OFF is the second operation associated with the [ON] key.) The grapher will turn itself off automatically after about five minutes without any activity.

It will be helpful to read the Getting Started section (pages 1 - 22) of the TI-85 Guidebook before proceeding.

Press [2nd] [MODE] to display the MODE settings. (MODE is the second operation associated with the [MORE] key.) Initially you should select the settings on the left side of the display.

To change a setting use [▽] or [△] to move the blinking cursor to the line of that setting. Then use [▷] or [◁] to move the cursor to the desired setting and press [ENTER]. Press [CLEAR] or [2nd] [QUIT] to leave the MODE screen. (QUIT is the second operation associated with the [EXIT] key.)

Chapter G
Introduction to Graphs and Graphers

SETTING THE VIEWING WINDOW

The viewing window is the portion of the coordinate plane that appears on the grapher's screen. It is defined by the minimum and maximum values of x and y: xMin, xMax, yMin, and yMax. The notation [Xmin, Xmax, Ymin, Ymax] is used in the text to represent these window settings or dimensions. For example, [−12, 12, −8, 8] denotes a window that displays the portion of the x-axis from −12 to 12 and the portion of the y-axis from −8 to 8. In addition, the distance between tick marks on the axes is defined by the settings xScl and yScl. The window corresponding to the settings [−20, 30, −12, 20], xScl = 5, yScl = 2, is shown below.

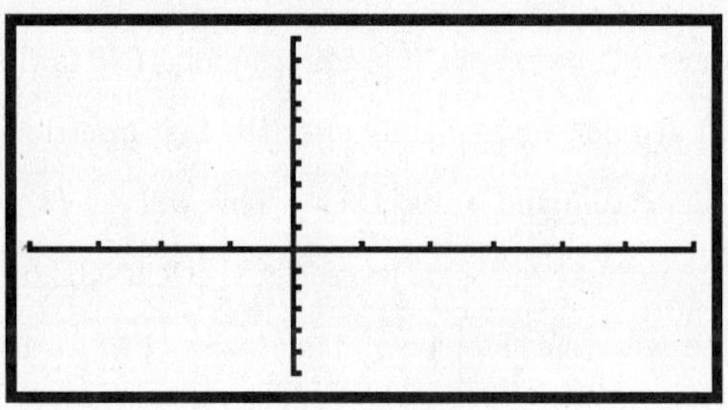

Press the GRAPH key in the left column of the keypad to display one level of GRAPH menus. Press F2 to select the RANGE menu which displays the current window settings on your grapher. The standard settings are shown below.

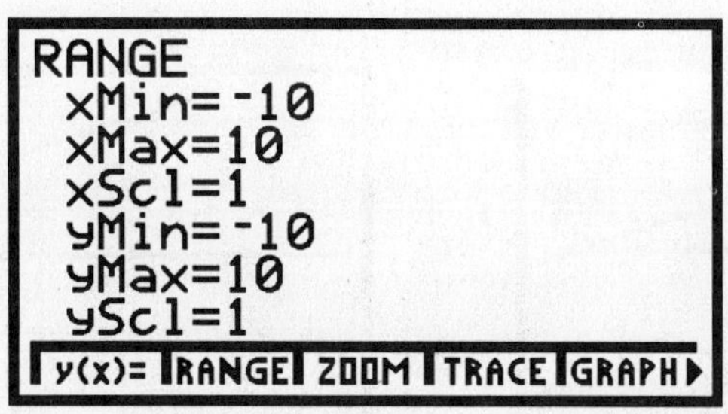

To change a setting press ▽ to move the cursor to the setting you wish to change and enter the new value. For example, to change from the standard settings to [−20, 30, −12, 20], xScl = 5, yScl = 2, press (−) 2 0 ENTER 3 0 ENTER 5 ENTER (−) 1 2 ENTER 2 0 ENTER 2 ENTER. You must use the gray (−) key on the bottom row of the keypad rather than the black − key in the right-hand column to enter a negative number. (−) represents "the opposite of" or "the additive inverse of" whereas − is the subtraction key. The ▽ key may be used instead of ENTER after typing each window setting.

To return quickly to the standard window setting [−10, 10, −10, 10], xScl = 1, yScl = 1, select the operations ZOOM and ZSTD from the GRAPH menu by pressing F3 followed by F4. If the GRAPH menu is not displayed, it will be necessary to press GRAPH first.

GRAPHING EQUATIONS

An equation must be solved for y before it can be graphed on the TI-85.

Example 8 (a), page 8 (Page numbers refer to pages in the textbook.): To graph $2x + 3y = 18$, first solve for y, obtaining $y = \dfrac{18 - 2x}{3}$. Then press GRAPH and select the menu y(x) = by pressing F1. If there is currently an expression displayed for $y1$, press CLEAR to delete it. Do the same for expressions that appear on all other lines by using ▽ to move the cursor to a line and then pressing CLEAR. Then use △ to move the cursor to the top line beside "$y1$ =." Now press (18 − 2 x-VAR) ÷ 3 to enter the right-hand side of the equation. Note that without the parentheses the expression $18 - \dfrac{2x}{3}$ would have been entered.

You can edit your entry if necessary. If, for instance, you pressed 5 instead of 8, use the ◁ key to move the cursor to 5 and then press 8 to overwrite it. If you forgot to type the right parenthesis, move the cursor to the division symbol /; then press 2nd INS) to insert the parenthesis before the division symbol. (INS is the second operation associated with the DEL key.) You can continue to insert symbols immediately after the first insertion without pressing 2nd INS again. If you typed 25 instead of 2, move the cursor to 5 and press DEL. This will delete the 5.

Once the equation is entered correctly, press 2nd M2 to select the RANGE menu and then select a viewing window. (M2 is the second operation associated with the F2 key.) Then press F5 to display the graph. You may change the viewing window as desired to reveal more or less of the graph. The standard window is shown here. The menus at the bottom of the window can be removed from the graph by pressing CLEAR. They will reappear when either EXIT or GRAPH is pressed.

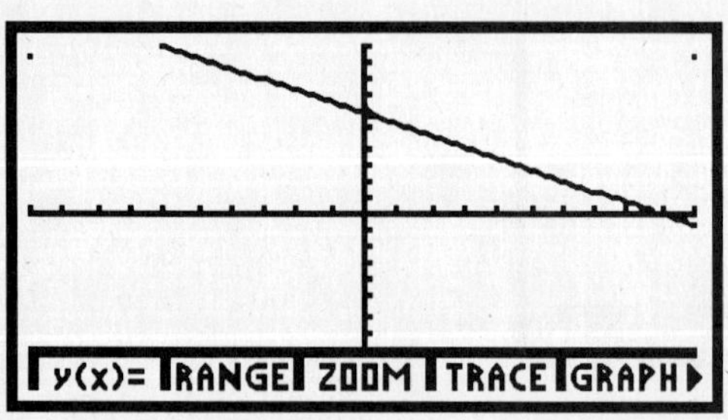

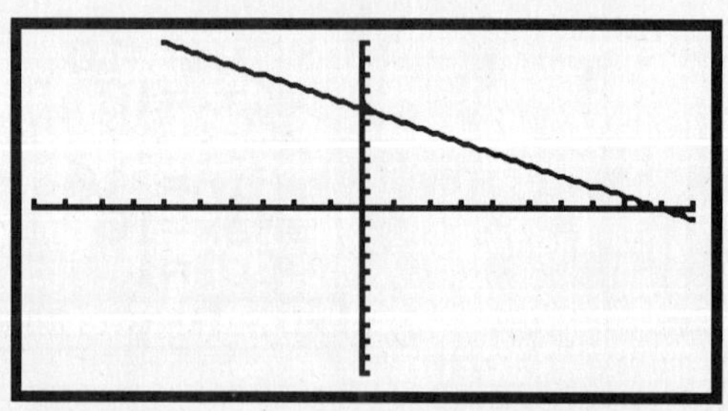

Example 8(c), page 8: To graph $x = y^2 + 1$, first solve the equation for y : $y = \pm\sqrt{x-1}$. To obtain the entire graph of $x = y^2 + 1$, you must graph $y_1 = \sqrt{x-1}$ and $y_2 = -\sqrt{x-1}$ on the same screen. Press GRAPH F1 to select the y(x) = menu and clear any expressions that currently appear. With the cursor beside "$y1$ =" press 2nd √ (x-VAR − 1). (√ is the second operation associated with the x^2 key.) Note that if the parentheses had not been used, the equation entered would have been $y_1 = \sqrt{x} - 1$.

Now use ▽ to move the cursor beside "$y2$ =." If "$y2$ =" is not visible, press ▽ and it will appear. There are two ways to enter $y_2 = -\sqrt{x-1}$. One way is to enter the expression $-\sqrt{x-1}$ directly by pressing (−) 2nd √ (x-VAR −

1).

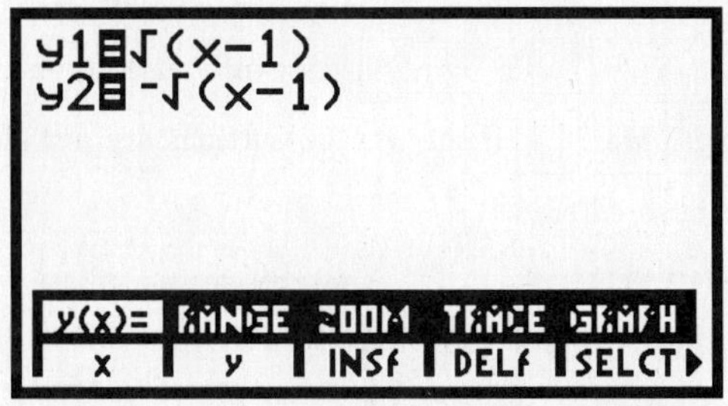

The other method of entering y_2 is based on the observation that $-\sqrt{x-1}$ is the opposite of the expression for y_1. That is, $y_2 = -y_1$. To enter this, place the cursor beside "$y2$ =" and press (−) F2 1. This enters the opposite of y_1 as the expression for y_2.

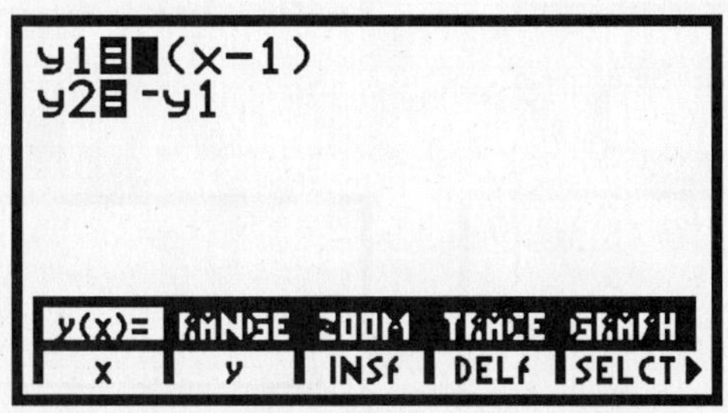

Press 2nd M2 to select the RANGE menu. Enter a viewing window and press F5 to display the graph. The window shown here is $[-2, 5, -5, 5]$, xScl = 1, yScl = 1.

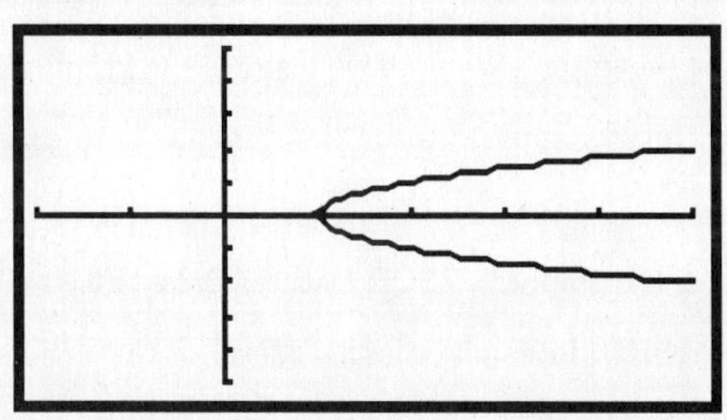

The top half is the graph of y_1, the bottom half is the graph of y_2, and together they yield the graph of $x = y^2 + 1$.

SQUARING THE VIEWING WINDOW

In the standard window, the distance between tick marks on the y-axis is about 7/12 the distance between tick marks on the x-axis. It is often desirable to choose window dimensions for which these distances are the same, creating a "square" window. Any window in which the ratio of the length of the y-axis to the length of the x-axis is 7/12 will produce this effect. (The actual ratio of the distances is 73/124 which is a little closer to 10/17 than to 7/12, but the last ratio is easier to work with than either of the other two and still gives a good representation.)

The desired ratio can be produced by selecting dimensions for which yMax − yMin $= \frac{7}{12}$(xMax − xMin). For example, the windows $[-12, 12, -7, 7]$ and $[-24, 24, -14, 14]$ are square. To illustrate this, graph the circle $x^2 + y^2 = 25$ in the standard window by first entering $y_1 = \sqrt{25 - x^2}$ and $y_2 = -\sqrt{25 - x^2}$ or $y_2 = -y_1$. Note that x^2 can be entered either by pressing [x-VAR] [x^2] or by pressing [x-VAR] [∧] 2. The [∧] key can be used to enter any exponent, but for the exponent 2 the [x^2] is more efficient. Press [2nd] [M3] [F4] if the window settings are not already standard or press [2nd] [M5] if the standard settings have previously been entered.

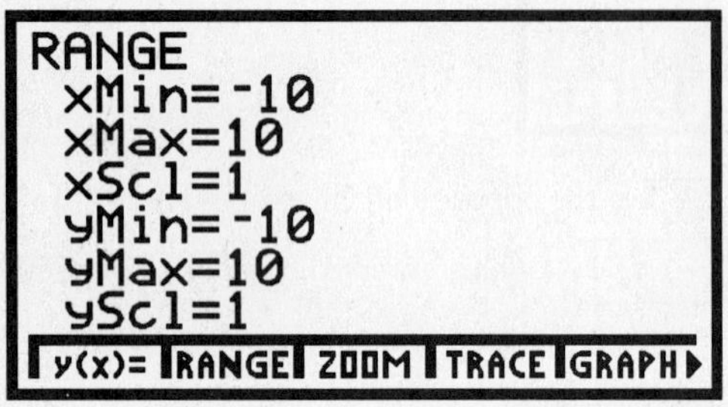

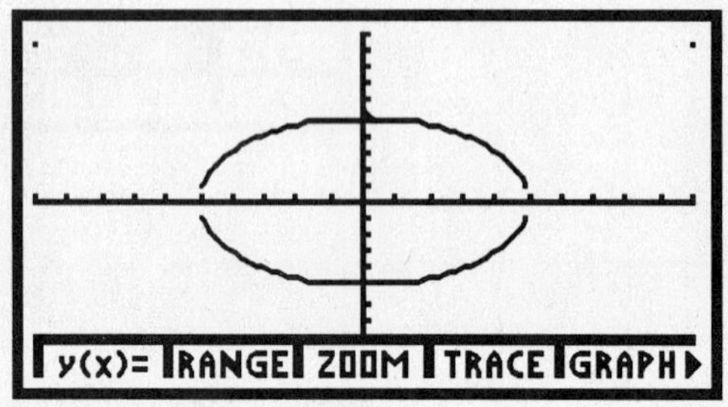

Note that the graph does not appear to be a circle.

Now change the window dimensions to $[-12, 12, -7, 7]$, xScl = 1, yScl = 1, and press [F5].

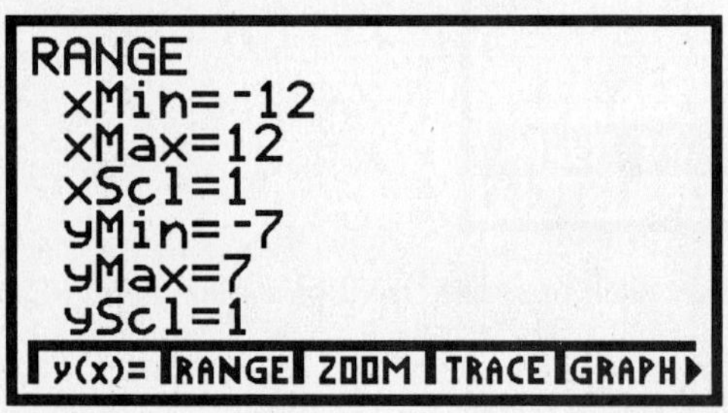

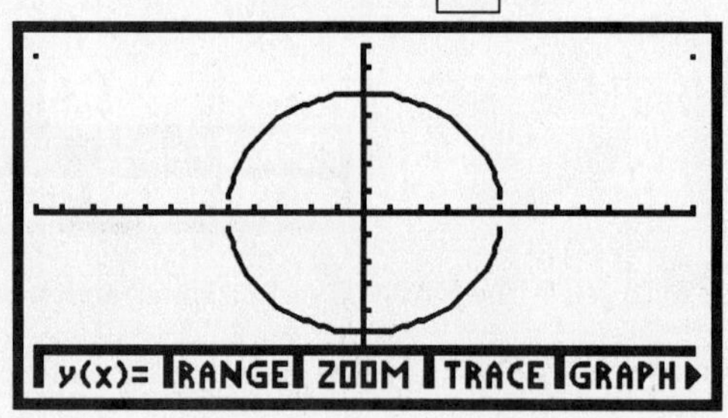

Although the software used to produce the graph shown above distorts the graph slightly, on your grapher screen you can observe that the distance between tick marks appears to be the same on both axes and that the graph appears to be a circle.

The window can also be squared using the grapher's ZSQR feature. Press [F3] [F4] to return to the graph of $x^2 + y^2 = 25$ in the standard window. Now press [MORE] to display the next level of the ZOOM menus and then press [F2] to select ZSQR. The resulting window dimensions and graph are shown below. Again, the software used to produce the graph introduces some distortion, but your grapher screen will show a graph that appears to be a circle.

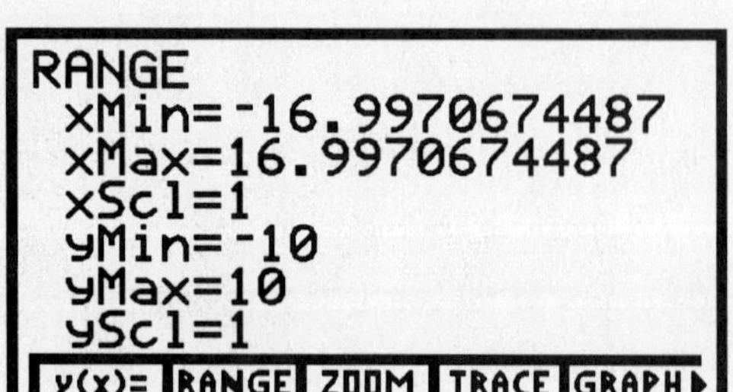

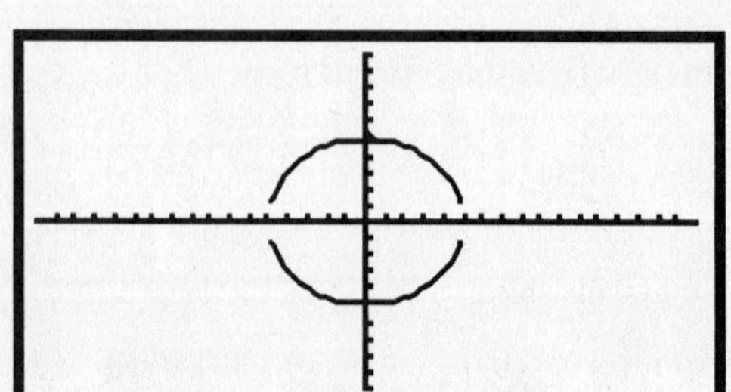

IDENTITIES

An equation that is true for every possible real-number substitution for the variable is an identity. The grapher can be used to provide a partial check whether an equation is an identity.

Example 9 (a), page 11: Determine whether $(x^2)^3 = x^6$ appears to be an identity.

To determine whether this equation appears to be an identity, graph $y_1 = (x^2)^3$ and $y_2 = x^6$. Examine the graphs in several viewing windows. The graphs appear to coincide no matter what the window. Thus, although there is a possibility that the graphs fail to coincide outside the windows that were examined, the equation appears to be an identity.

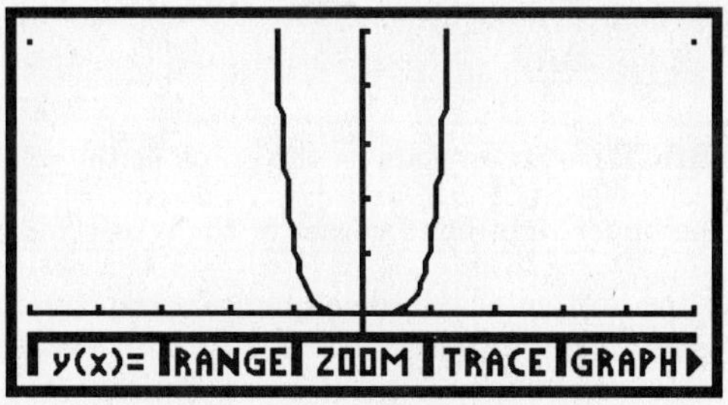

Example 9 (b), page 11: Determine whether $\sqrt{x+4} = \sqrt{x} + 2$ appears to be an identity.

To determine whether this equation appears to be an identity, graph $y_1 = \sqrt{x+4}$ and $y_2 = \sqrt{x} + 2$. Note that parentheses must be used when entering $\sqrt{x+4}$: [2nd] [$\sqrt{\ }$] [(] [x-VAR] [+] [4] [)]. Without parentheses the expression entered would be $\sqrt{x} + 4$.

Any window that includes a portion of the first quadrant will show that the graphs differ. Thus, the equation is not an identity.

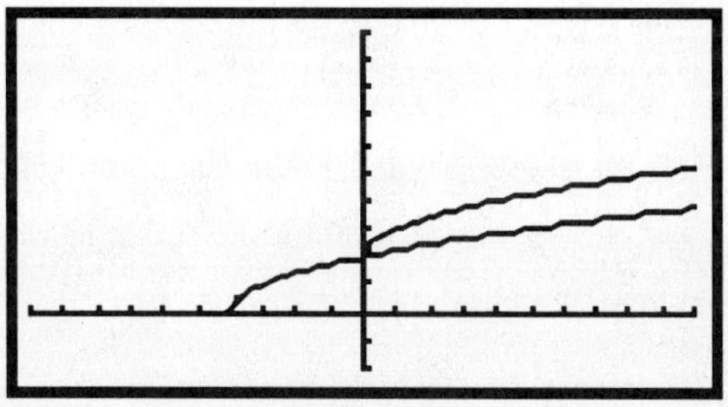

SOLVING EQUATIONS USING TRACE AND ZOOM

There are several techniques that can be used to solve equations with a grapher. One such technique uses the grapher's TRACE and ZOOM features.

Example 10, page 13: Solve $x^3 - 3x + 1 = 0$. Approximate the solutions to three decimal places.

The solutions of this equation are the first coordinates of the x-intercepts of the graph of $y = x^3 - 3x + 1$. To find these coordinates we first graph $y = x^3 - 3x + 1$ in a viewing window that shows all of the x-intercepts. Here we use the standard window, $[-10, 10, -10, 10]$.

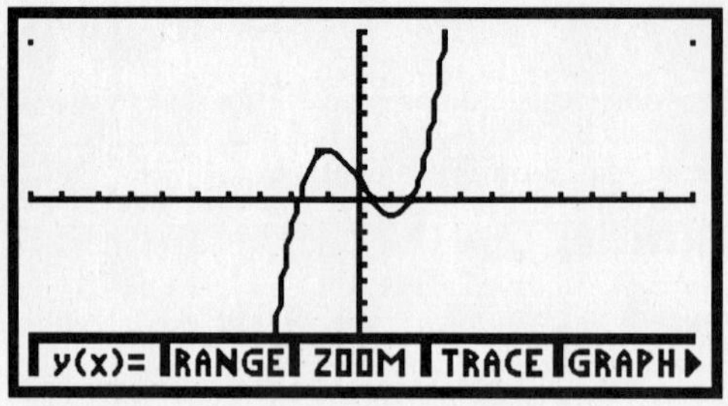

We see that x-intercepts occur near $x = -2$, $x = 0$, and $x = 2$. A portion of the viewing window can be enlarged near each of these values in order to find the desired three decimal place approximation. For example, let's examine the graph near $x = 0$.

Press F4 to select TRACE. The TRACE cursor appears on the graph at the middle x-value of the window, in this case at $x = 0$. The number 1 appears in the upper right-hand corner of the screen indicating that the cursor is on the graph of equation y_1. The x-and y-values at the bottom of the screen indicate the coordinates of the point where the cursor is positioned, in this case at $x = 0$, $y = 1$.

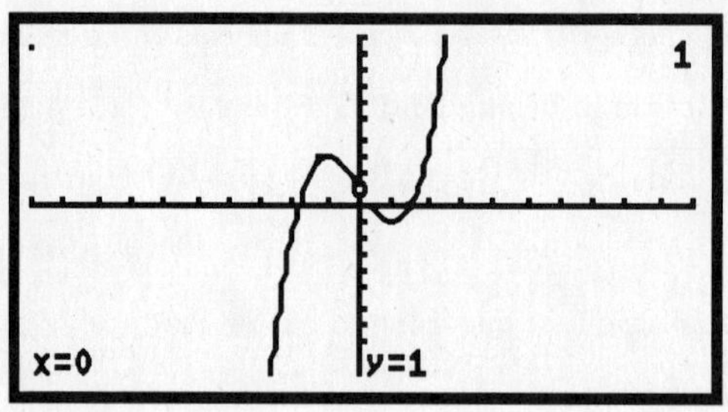

Pressing ◁ or ▷ moves the cursor to the left or right along the curve. Note that the TRACE cursor always remains on the curve.

In order to find the middle x-intercept, we enlarge the portion of the graph near $x = 0$ by first positioning the cursor as close to this intercept as possible. That is, near $x = 0$, position the cursor as close as possible to $y = 0$. Press EXIT. Then press F3 F2 ENTER to zoom in on the graph near $x = 0$.

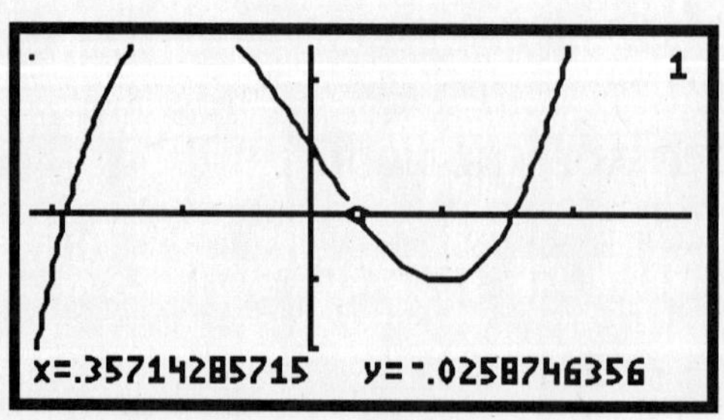

Press GRAPH F4 to select TRACE again and move the cursor as close as possible to $y = 0$ near $x = 0$. Now press EXIT F3 F2 ENTER again to enlarge this portion of the graph further.

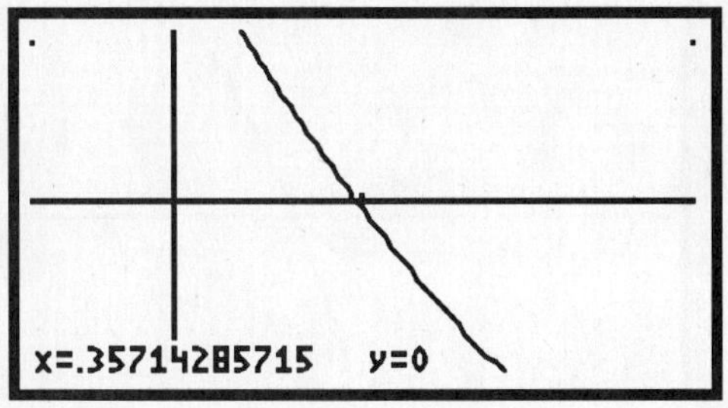

Continue tracing and zooming in until the x-values determined by positioning the cursor just to the left and just to the right of the x-intercept are the same when rounded to three decimal places. We find that $x \approx 0.347$ at the middle x-intercept. This is one solution of the equation $x^3 - 3x + 1 = 0$.

To find a second solution of the equation press GRAPH F3 F4 to return to the standard window; then trace to a position on the curve near another x-intercept and zoom in as described above. Continue to trace and zoom in until the desired accuracy is obtained. Repeat this process to find the third solution. The other two solutions are about -1.879 and 1.532.

At any point in this process, we can zoom out to the previous window by pressing GRAPH F3 F3 ENTER. This can be repeated as many times as desired.

The ZOOM BOX operation can also be used to enlarge a portion of the viewing window. We select diagonally opposed corners of a box that defines the new window. For example, graph $y = x^3 - 3x + 1$ in the standard window. Either one or two rows of menus will appear at the bottom of the screen, depending on the order in which the graph and the window were chosen. If one row appears, press F3 F1 to access the BOX operation. If there are two rows, press F1 to select Box. Then move the cursor from the center of the screen to any corner of the box to be defined. Note that the cursor used in this operation is a free-moving cursor. That is, it can be positioned at any point on the screen as opposed to the trace cursor whose position is restricted to points on the graph.

Press ENTER and a small square dot appears, indicating that the first corner has been selected. Now use ◁, ▷, △, and/or ▽ to move the cursor to the corner of the box diagonally opposite this one. As you move the cursor away from the first corner the boundaries of the box appear on the screen and change as the position of the cursor changes.

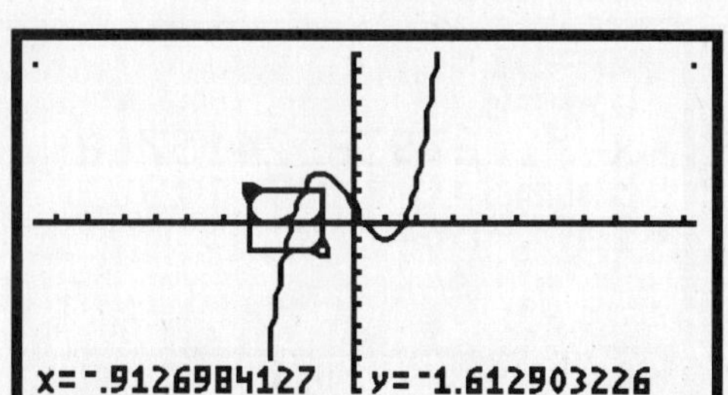

When the box is defined as you want it, press ENTER to show only the portion of the graph in the box.

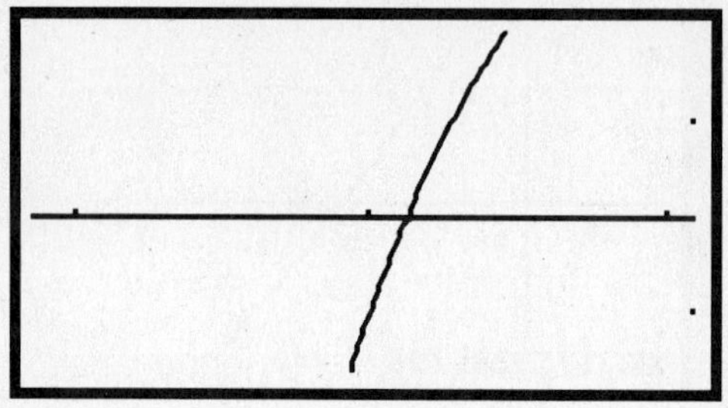

Now we can zoom in or use ZOOM BOX again to further enlarge a portion of the graph.

SOLVING EQUATIONS USING SOLVER

The SOLVER operation can be used to solve equations. An equation need not be written with 0 on one side of the equals sign when this operation is used.

Example 10, page 13: Solve $x^3 - 3x + 1 = 0$. Approximate the solutions to three decimal places.

Begin as in the procedure described above for using TRACE and ZOOM by graphing $y = x^3 - 3x + 1$ in a window that shows all of the x-intercepts. Note again that the x-intercepts occur near $x = -2$, $x = 0$, and $x = 2$. We will first find the solution near $x = -2$.

Access the SOLVER operation by pressing 2nd SOLVER. (SOLVER is the second operation associated with the GRAPH key.) The notation "eqn:" appears on the screen with the blinking cursor positioned to the right of the colon. If an expression appears on this line, delete it by pressing CLEAR. Then since the equation has already been entered on the "y(x) =" screen as $y1$, we can enter it here by pressing F4 ENTER. It can also be entered by pressing x-VAR ∧ 3 − 3 x-VAR + 1 ENTER. Note that we have entered the expression $x^3 - 3x + 1$ rather than the equation $x^3 - 3x + 1 = 0$. The SOLVER edit screen now appears with the blinking cursor positioned to the right of the notation "exp =." Since we are seeking the values of x for which the expression $x^3 - 3x + 1$ is equal to 0, type 0 ENTER here. The cursor now appears beside "x =," where a guess for the solution should be entered. Since we are seeking the solution near $x = -2$, type (−) 2 here. Press F5 to select the SOLVE operation. The grapher returns the value −1.8793852415718. Thus one solution is about −1.879. Note that the cursor must be positioned on the "x =" line when SOLVE is selected.

To find the solution near $x = 0$ clear the solution that was just found. This is the entry beside "x =." Now enter 0 on this line and press F5 to solve again. The grapher returns the value .34729635533385. To find the solution near $x = 2$, repeat the previous process, entering 2 on the "$x =$" line. The value 1.5320888862379 is returned. Thus, as we saw before, the other two solutions are about 0.347 and 1.532.

Example 12, page 15: Solve: $\frac{2}{3}x - 7 = 5$.

The most direct option is to graph is to graph $y_1 = \frac{2}{3}x - 7$ and $y_2 = 5$ in a window that shows the point of intersection of the graphs. Here we use $[-5, 30, -10, 10]$, xScl = 5, yScl = 1.

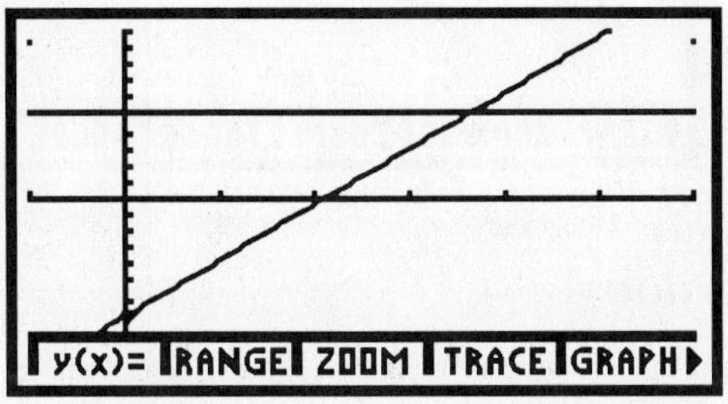

Now press 2nd SOLVER to access the SOLVER operation. Since $\frac{2}{3}x - 7$ and 5 have already been entered in the "y(x) =" screen as $y1$ and $y2$, respectively, enter $y1 = y2$ beside "eqn:" by pressing F4 ALPHA = F5. (= is the ALPHA operation associated with the STO▷ key.) We could also enter $\frac{2}{3}x - 7 = 5$ directly by pressing (2 ÷ 3) x-VAR − 7 ALPHA = 5. Now press enter and the cursor appears beside "$x =$," where a guess for the solution should be entered. The graph shows that the point of intersection is near $x = 20$, so we enter 20 for the guess. Then, with the cursor on the "$x =$" line, press F5 to select SOLVE. The grapher returns the solution, 18.

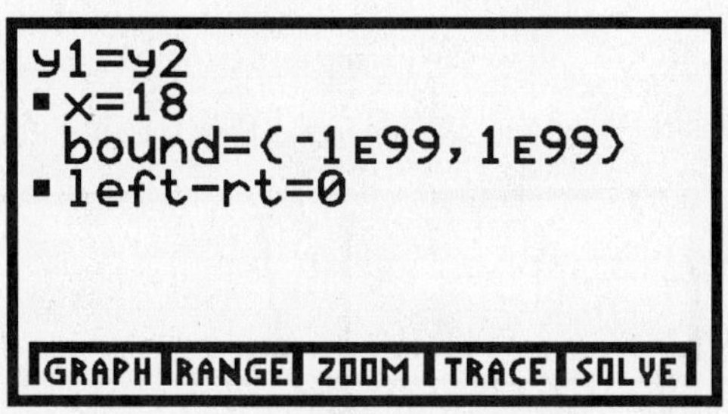

Another option is to write the equation with 0 on one side of the equals sign: $\frac{2}{3}x - 7 - 5 = 0$, or $\frac{2}{3}x - 12 = 0$. Then graph $y = \frac{2}{3}x - 12$ in a window that shows the x-intercept. Here we use $[-5, 35, -10, 10]$, xScl = 5, yScl = 2.

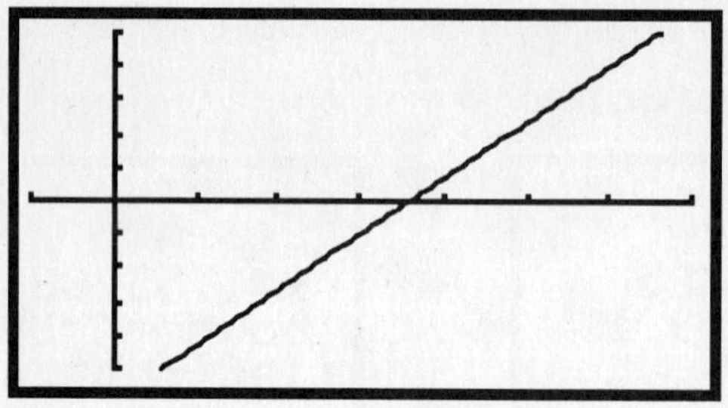

Repeat the procedure of Example 10 above. Since the x-intercept is near $x = 20$, use 20 as the guess for the solution on the "$x =$" line. The grapher returns the solution of the equation, 18.

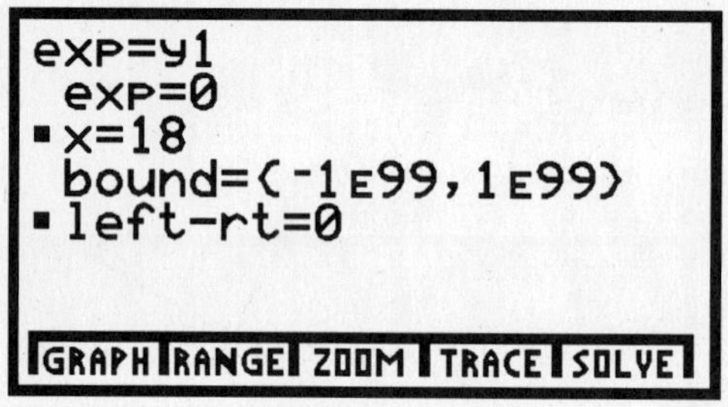

SOLVING EQUATIONS USING ROOT

A solution of an equation with zero on one side is usually called a zero, but is sometimes called a root. The ROOT operation can also be used to solve an equation.

Example 10, page 13: Solve $x^3 - 3x + 1 = 0$. Approximate the solutions to three decimal places.

Graph $y = x^3 - 3x + 1$ in the standard window as before. We will use ROOT to find the solution of $x^3 - 3x + 1 = 0$ near $x = 0$. Access the ROOT operation by pressing MORE F1 to select the MATH menu. Now the choices LOWER, UPPER, and ROOT appear on the screen below the graph. Press F1 to select LOWER. Position the cursor just to the left of the x-intercept near 0 and press ENTER. Then press F2 to select UPPER. Position the cursor just to the right of the intercept and press ENTER again. The upper and lower bounds are indicated by two delineators shown near the top of the screen. Now press F3 to select ROOT. Move the cursor as close to the x-intercept as possible and press ENTER. The x-value that the grapher returns is the solution. We see again that the solution is about 0.347.

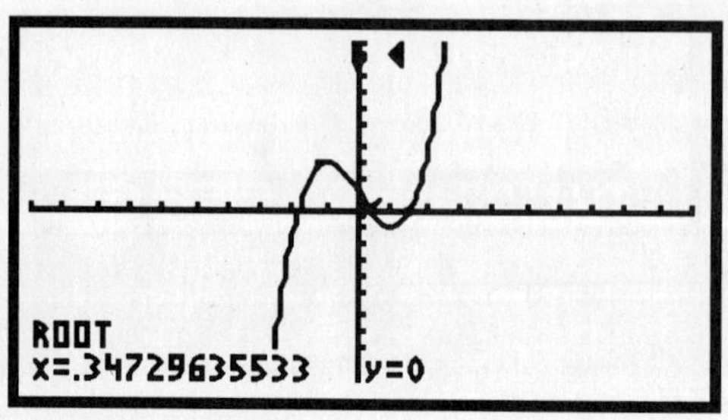

The other two solutions of the equation can be found by using the ROOT operation two more times.

FINDING POINTS OF INTERSECTION

There are several ways in which the grapher can be used to determine the point(s) of intersection of two graphs.

Example 11, page 14: Find the point of intersection of the graphs of the equations $y_1 = 3x^5 - 20x^3$ and $y_2 = 34.7 - 1.28x^2$. Approximate the coordinates to three decimal places.

First graph the equations in a window that shows all of the points of intersection. We use $[-4, 4, -80, 80]$, xScl = 1, yScl = 20 here.

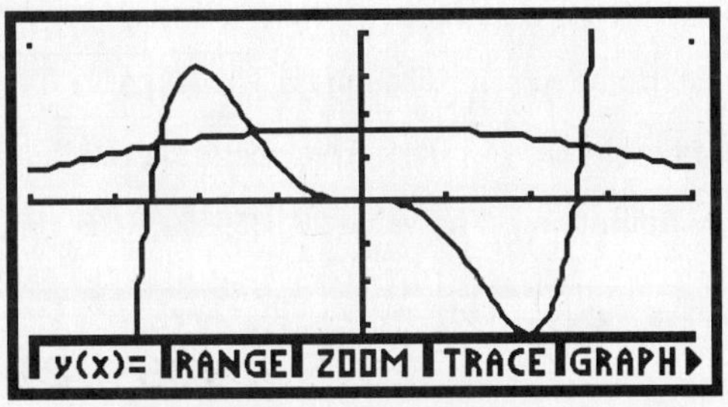

There are three points of intersection. One way to find their coordinates is to use TRACE and ZOOM. (See "Solving Equations Using Trace and Zoom" on page 91 of this manual.) Another method is to use the ISECT operation. We will illustrate its use by finding the point of intersection at the far left. After graphing y_1 and y_2 as above, access ISECT by pressing MORE F1 to select MATH. Set the lower and upper bounds as described in the previous example. Then press MORE F5 to select ISECT. The cursor appears on the graph of y_1. Press ENTER. The cursor moves to the next function, y_2. Move the cursor as close as possible to the point of intersection and press ENTER again. The grapher returns the coordinates of the point of intersection. We see that they are about $(-2.463, 26.936)$.

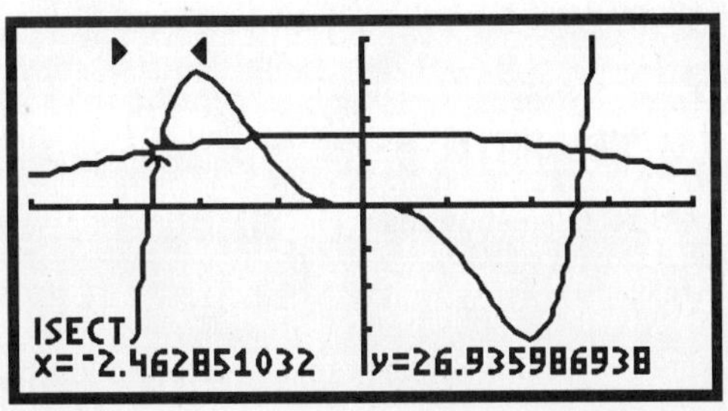

The coordinates of the other two points of intersection can be found by using the ISECT operation two more times.

The SOLVER feature can also be used to find the x-coordinates of the points of intersection by finding solutions of $y_1 = y_2$. Again it is necessary to graph y_1 and y_2 in order to find guesses for the solutions.

To find the point of intersection on the far left, first note that its x-coordinate is about -2.5. This is the value we will use for the guess. Press 2nd SOLVER then F4 ALPHA = F5 ENTER to enter $y_1 = y_2$ beside "eqn:." Now type -2.5 following "x =." Press F5 to select SOLVE. The grapher returns the value -2.4628510317595. This is the first coordinate of the point of intersection.

This value must be substituted into either y_1 or y_2 to find the second coordinate. To substitute in y_1 after using SOLVE first press [2nd] [QUIT] to return to the Home screen. Then press 3 [2nd] [ANS] [∧] 5 [−] 20 [2nd] [ANS] [∧] 3 [ENTER]. (ANS is the second operation associated with the [(−)] key. This operation recalls the answer to the previous computation, thus allowing it to be used in a new computation.) The value 26.9359869381 is returned.

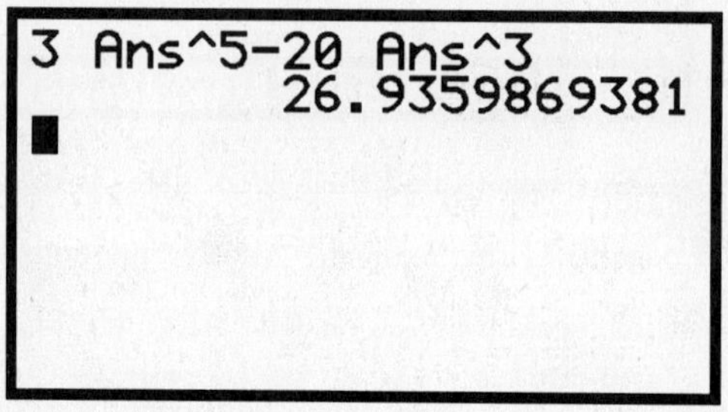

Thus, we see that the coordinates of the point of intersection are about $(-2.463, 26.936)$. If we had used the rounded x-value of -2.463 in y_1 to compute the corresponding y-value, the result would have been 26.908. If we had used the rounded x-value in y_2, the resulting y-value would have been 26.935. To achieve the greatest accuracy, then, it is important to use the unrounded x-value in this computation.

SOLVING EQUATIONS USING INTERSECT

The ISECT operation can also be used to solve equations.

Example 12, page 15: Solve: $\frac{2}{3}x - 7 = 5$.

The solution of this equation is the first coordinate of the point of intersection of the graphs of $y_1 = \frac{2}{3}x - 7$ and $y_2 = 5$. Thus, we graph these equations in a window that shows the point of intersection and use ISECT as described in "Finding Points of Intersection" on page 97 of this manual. The x-coordinate of the point of intersection is 18. This is the solution of the equation.

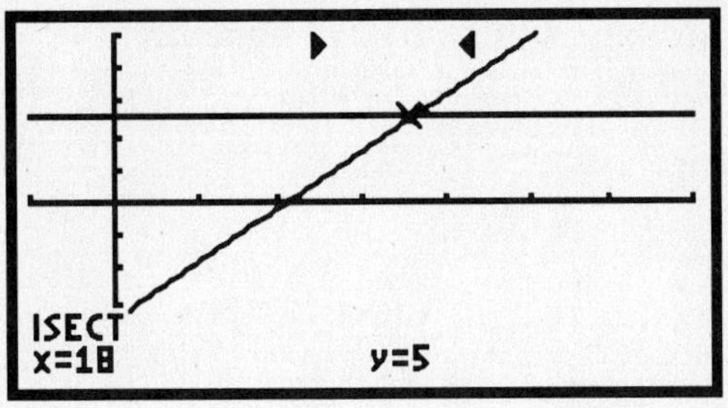

SELECTING DOT MODE

When graphing an equation in which a variable appears in a denominator, DrawDot mode should be used. To select this mode press GRAPH MORE F3 to select FORMT (format) and use ▽ and ▷ to position the blinking cursor over "DrawDot." Then press ENTER.

See page 16 in the text for further explanation of why DrawDot mode is used in certain cases and for an illustration of an equation graphed in both DrawLine (or CONNECTED) and DrawDot (or DOT) modes.

Chapter 1
Graphs, Functions, and Models

FINDING FUNCTION VALUES

When a formula for a function is given, function values for real-numbered inputs can be found in several ways on a grapher.

Example 4 (a), (b), page 23 (Page numbers refer to pages in the textbook.): For $f(x) = 2x^2 - x + 3$, find $f(0)$ and $f(-7)$.

Method 1: Substitute the inputs directly in the formula. For example, to find $f(0)$ press 2 × 0 x^2 − 0 + 3 ENTER. To find $f(-7)$ press 2 ((−) 7) x^2 − ((−) 7) + 3 ENTER.

```
2*0²-0+3
                    3
2(-7)²-(-7)+3
                  108
█
```

Method 2: Enter $y_1 = 2x^2 - x + 3$ in the "y(x) =" screen. Then press 2nd QUIT to go to the home screen. (QUIT is the second operation associated with the EXIT key.) Now to find $f(0)$, the value of y_1 when $x = 0$, press 0 STO ▷ x-VAR 2nd : 2nd VARS MORE F3 to select EQU. Then use ▽ to move the triangular selection indicator beside $y1$ and press ENTER ENTER. (: is the second operation associated with the · key. VARS is the second operation associated with the 3 key.) This series of keystrokes stores 0 as the value of x and then substitutes it in the function y_1. To find the function value at $x = -7$, repeat this process, using −7 in place of 0 or press 2nd ENTRY to recall the previous entry, move the cursor to 0, and replace 0 with −7. Note that since two keystrokes are required to enter −7, it is necessary to use 2nd INS to change 0 to −7. That is, with the cursor on 0 press (−) 2nd INS 7 ENTER to make this change.

```
0→x:y1
                    3
-7→x:y1
                  108
```

Method 3: Enter $y_1 = 2x^2 - x + 3$ in the "y(x) =" screen and press 2nd QUIT to go to the home screen. Then, to find

$f(0)$ press [2nd] [MATH] [F5] to select MISC. (MATH is the second operation associated with the [×] key.) Then press [MORE] [F5] to select "eval" followed by 0 [ENTER]. To find $f(-7)$ either repeat this process entering -7 in place of 0 or recall the first entry and replace 0 with -7 and press [ENTER].

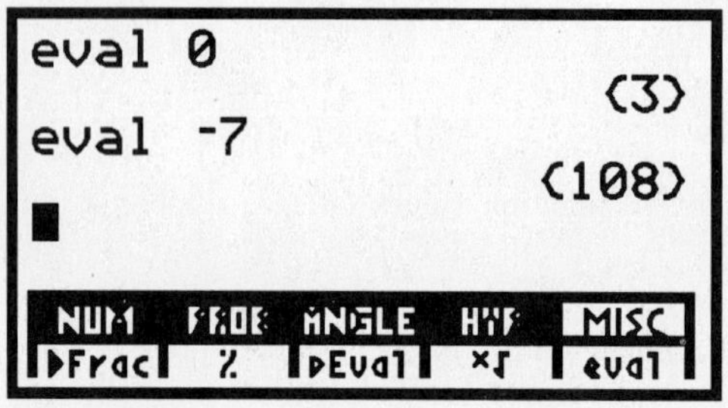

GRAPHING FUNCTIONS DEFINED PIECEWISE

Operations from the TEST menu are used to enter functions that are defined piecewise. (TEST is the second operation associated with the [2] key.) Select DrawDot mode from the FORMT menu after pressing [GRAPH] [MORE] when graphing such functions.

Example 4, page 38: Graph

$$f(x) = \begin{cases} 4, & \text{for } x \le 0, \\ 4 - x^2, & \text{for } 0 < x \le 2, \\ 2x - 6, & \text{for } x > 2, \end{cases}$$

Press [GRAPH] [F1] to select y(x) = and clear any functions that have previously been entered. With the cursor beside "$y1$ =" enter the function as described in the middle of page 102 of the text by pressing [(] 4 [)] [(] [x-VAR] [2nd] [TEST] [F4] 0 [)] [+] [(] 4 [−] [x-VAR] [x^2] [)] [(] 0 [F2] [x-VAR] [)] [(] [x-VAR] [F4] 2 [)] [+] [(] 2 [x-VAR] [−] 6 [)] [(] [x-VAR] [F3] 2 [)]. Select a window and then press [GRAPH].

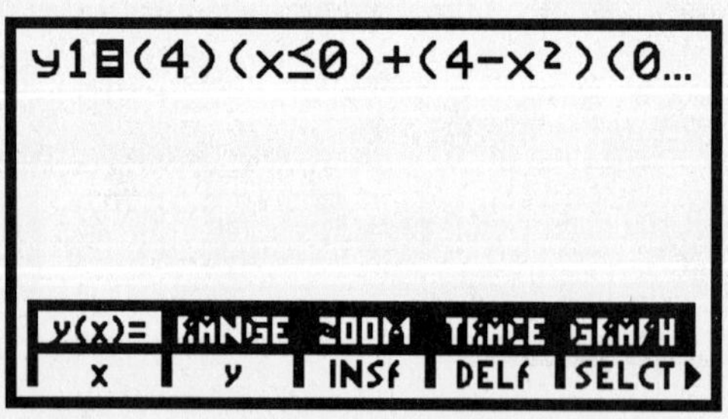

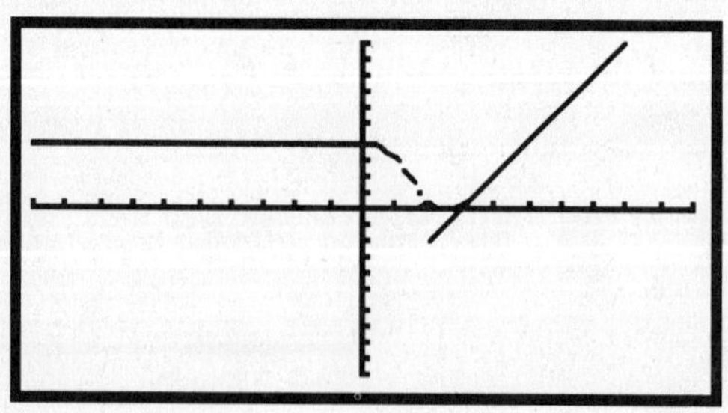

Example 6, page 39: Graph $f(x) = \text{INT}(x)$.

With the grapher set in DrawDot mode, press [GRAPH] [F1] to select y(x) = and clear any previously entered functions. Position the cursor beside "$y1$ =" and select the greatest integer function from the MATH NUM menu as follows. Press [2nd] [MATH] [F1] to select NUM. Then press [F4] to select int followed by [x-VAR]. Now press [EXIT] to return to the menu that allows you to select a window and graph the function.

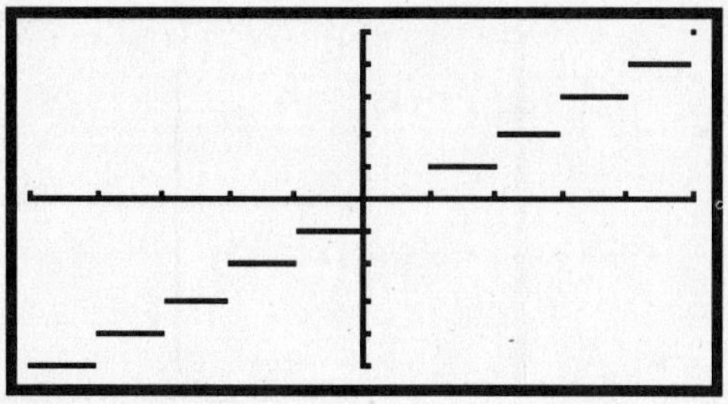

GRAPHING VERTICAL LINES

Although equations of vertical lines cannot be entered in the grapher in "y(x) =" form, vertical lines can be graphed using the DRAW feature or by typing an instruction on the home screen. DRAW is accessed by pressing [GRAPH] [MORE].

Interactive Discovery, top of page 56: Graph $x = -5$ and $x = 2$.

To graph $x = -5$ using the DRAW feature, first press [GRAPH] [F1] to select y(x) = and clear all previously entered equations. Then select a window. On the RANGE screen press [MORE] [F2] to select DRAW. Then press [F3] to select VERT. Use [◁] and [▷] to position the cursor where the line should be located and press [ENTER]. To graph $x = 2$ on the same screen use [▷] and [◁] to reposition the cursor and press [ENTER].

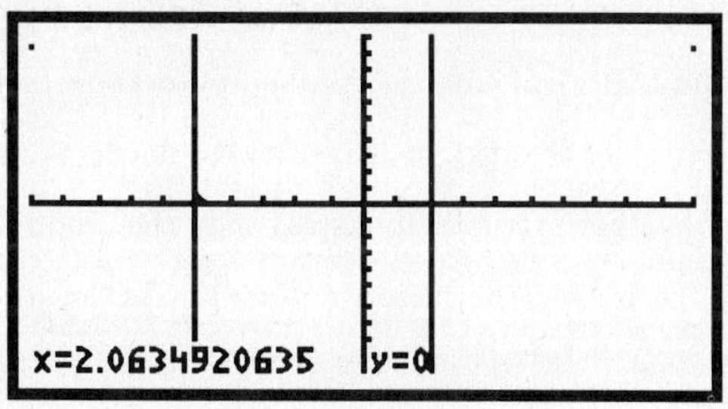

As the coordinates at the bottom of the window indicate, this method does not allow precise placement of vertical lines. Vertical lines can be placed precisely by drawing them from the home screen as described below. First clear the drawings shown above from the screen by pressing [GRAPH] [MORE] [F2] or [EXIT] [2nd] [M2] to access the DRAW menu. Then press [MORE] [F5] to select the CLDRW (clear drawing) operation.

To graph $x = -5$ from the home screen, first press [GRAPH] [F1] to select y(x) = and clear all previously entered equations. Then select a window. Now press [2nd] [QUIT] to go to the home screen. Press [ALPHA] [ALPHA] to lock in the uppercase ALPHA keys and type VERT; press [(−)] to provide a space, then [ALPHA] to unlock the uppercase ALPHA keys. Press [(−)] 5 [ENTER] to see the graph of $x = -5$. To graph $x = -5$ and $x = 2$ on the same screen, after pressing [ENTER] press [2nd] [QUIT] [2nd] [ENTRY]. Then edit this entry to change −5 to 2 and press [ENTER].

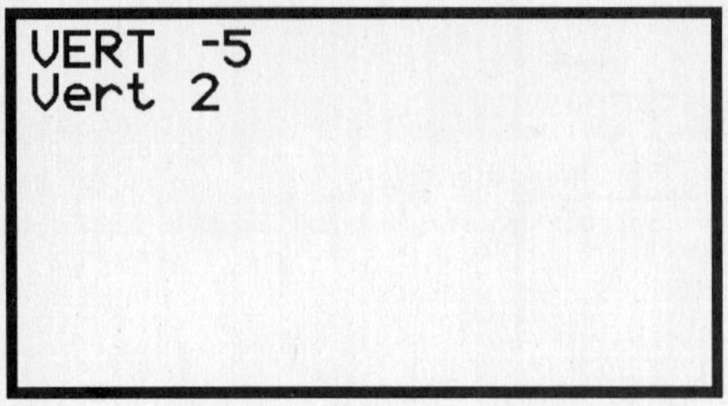

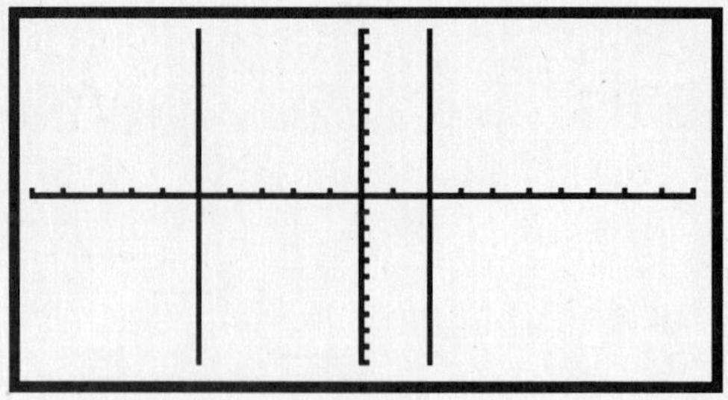

These lines can be cleared from the graph screen using the DRAW menu as described above. Whether graphed using DRAW or from the home screen, graphs of vertical lines will also be cleared when another function is subsequently entered on the "y(x) =" screen and graphed.

THE ALGEBRA OF FUNCTIONS

The grapher can be used to evaluate and graph combinations of functions.

Example 1 (b), page 92: Given that $f(x) = x + 1$ and $g(x) = \sqrt{x+3}$, find $(f + g)(6)$.

Press [GRAPH] [F1] to select y(x) = and enter $y_1 = x + 1$, $y_2 = \sqrt{x+3}$, and $y_3 = y_1 + y_2$. To enter $y_3 = y_1 + y_2$ press [F2] 1 [+] [F2] 2. Note that $y_3 = f(x) + g(x)$, or $(f + g)(x)$. Use y_3 to find $(f + g)(6)$ using one of the methods for finding function values described on page 81 of this manual. We find that $(f + g)(6) = 10$.

To view the graphs of $f(x)$, $g(x)$, and $(f + g)(x)$ enter y_1, y_2, and y_3 as above, select a window, and press [F5] to select GRAPH. These graphs appear on page 159 of the text. It is possible to deselect one or two of these functions and display the graph(s) of the remaining function(s). For example, to display only the graph of y_3 without deleting the equations of y_1 and y_2, press [F1] to select y(x) =. Then move the cursor to y_1 and press [F5] to select SELCT. This deselects or turns off y_1. Do the same for y_2. Now press [2nd] [M5] to select GRAPH and see only the graph of y_3.

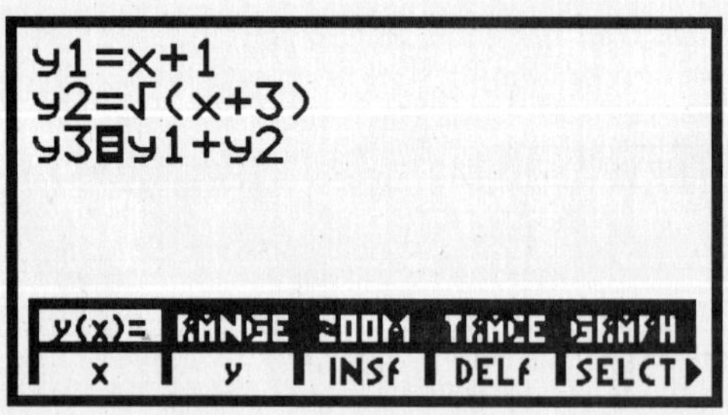

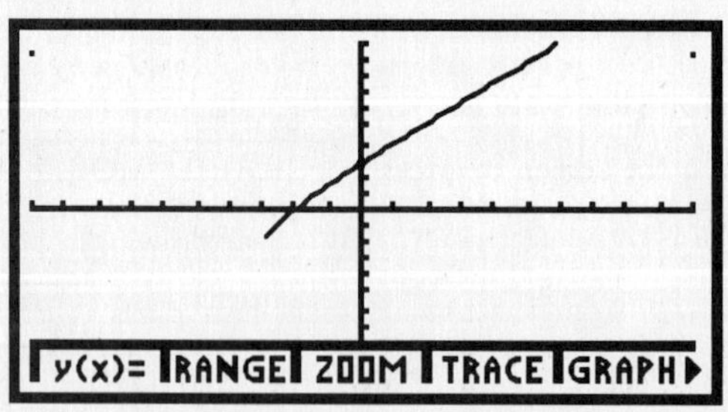

To select or turn on a function again, repeat this process. The SELCT operation reverses the selection status of a function. That is, it deselects a selected function and selects a deselected function. Note that the equals sign on a selected function is highlighted.

Example 3 (a), page 95: Given that $f(x) = 2x - 5$ and $g(x) = x^2 - 3x + 8$, find $(f \circ g)(7)$ and $(g \circ f)(7)$.

Press [GRAPH] [F1] to select y(x) = and enter $y_1 = 2x - 5$, $y_2 = x^2 - 3x + 8$, $y_3 = 2y_2 - 5$, and $y_4 = y_1^2 - 3y_1 + 8$. Use the y menu operation as described in Example 1(b) above to enter y_3 and y_4. Note that $y_3 = (f \circ g)(x)$ and $y_4 = (g \circ f)(x)$. Use

y_3 and y_4 to find $(f \circ g)(7)$ and $(g \circ f)(7)$, respectively, employing one of the methods for finding function values described on page 81 of this manual. Note that when we use the "eval" operation from the MATH menu with all four of the functions y_1, y_2, y_3, and y_4 selected, then the last two values displayed correspond to $y_3(7)$ and $y_4(7)$, respectively, or to $(f \circ g)(7)$ and $(g \circ f)(7)$. This situation is shown in the figure on the right below.

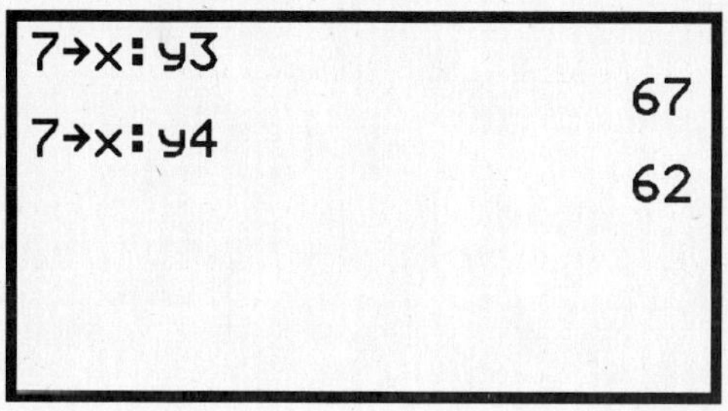

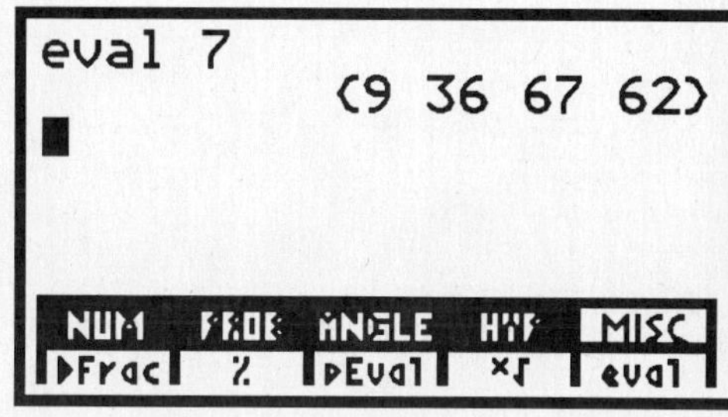

GRAPHING AN INVERSE FUNCTION

The DrawInv operation on the TI-85 can be used to graph a function and its inverse on the same screen. A formula for the inverse function need not be found in order to do this. The grapher must be set in Func mode when this operation is used.

Example 7, page 105: Graph $f(x) = 2x - 3$ and $f^{-1}(x)$ using the same set of axes.

Be sure the GRAPH FORMT is set to DrawLine and then enter $y_1 = 2x - 3$ and either clear or deselect all other functions on the "y(x) =" screen. Press [2nd] [M5] to select GRAPH. Then press [MORE] [F2] to select DRAW and [MORE] [MORE] [F2] to select the DrInv operation. Press [2nd] [alpha] and then type $y1$. Press [ENTER] to see the graph of the function and its inverse.

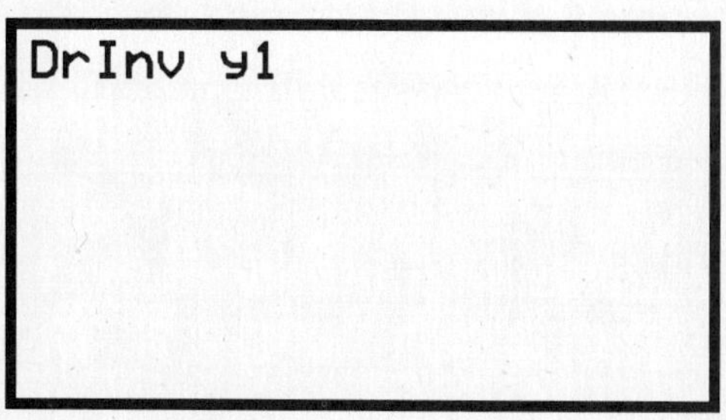

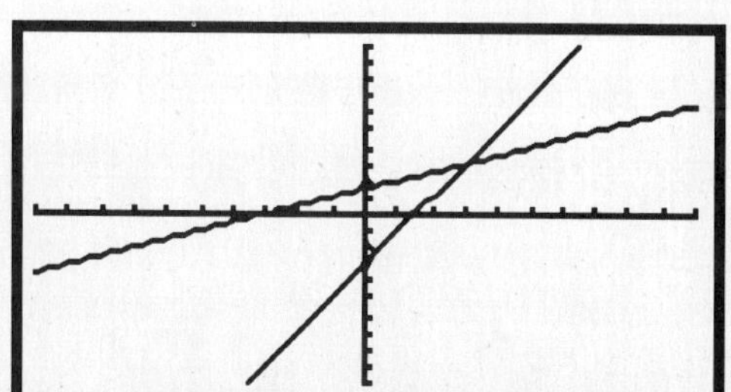

Chapter 2
The Trigonometric Functions

CONVERTING BETWEEN D°M'S" AND DECIMAL DEGREE MEASURE

The ANGLE feature can be used to convert D°M'S" notation to decimal notation and vice versa. ANGLE is accessed by pressing 2nd MATH F3.

Example 5, page 122: Convert 5°42′30″ to decimal degree notation.

Enter 5°42′30″ as 5′42′30′ by pressing 5 2nd MATH F3 5 F3 4 2 F3 3 0 F3 ENTER. The grapher returns 5.70833333333, so 5°42′30″ ≈ 5.71°.

Example 6, page 122: Convert 72.18°to D°M'S" notation.

Press 7 2 · 1 8 2nd MATH F3 to select ANGLE. Then press F4 to select DMS followed by ENTER. The grapher returns 72°10′48″.

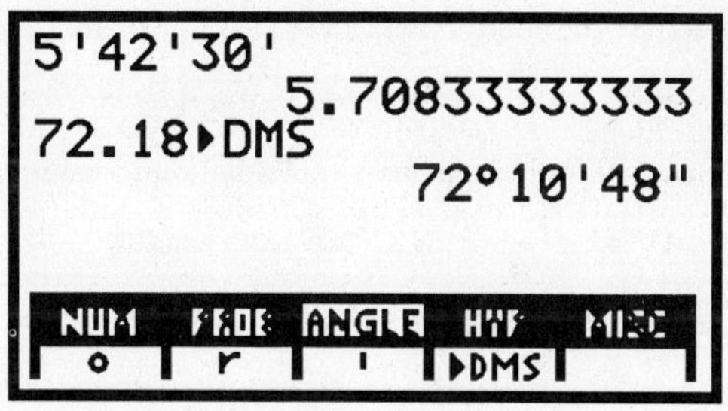

FINDING TRIGONOMETRIC FUNCTION VALUES

The grapher's SIN, COS, and TAN operations can be used to find the values of trigonometric functions. When angles are given in degree measure, the grapher must be set in Degree mode.

Example 7, page 122: Find the trigonometric function value, rounded to four decimal places, of each of the following.

a) tan 29.7° b) sec 48° c) sin 84°10′39″

a) Press TAN 2 9 · 7 ENTER. We find that tan 29.7° ≈ 0.5704.

b) The secant, cosecant, and cotangent functions can be found by taking the reciprocals of the cosine, sine, and tangent functions, respectively. This can be done either by entering the reciprocal or by using the x^{-1} key. To find sec 48° we can enter the reciprocal of cos 48° by pressing 1 ÷ COS 4 8 ENTER. To find sec 48° using the x^{-1} key press (COS 4 8) 2nd x^{-1} ENTER. (x^{-1} is the second operation associated with the EE key.) The result is sec 48° ≈ 1.4945.

```
1/cos 48
            1.49447654986
(cos 48)-1
            1.49447654986
■
```

c) Press [SIN] followed by 84°10′39″ entered as described above in Converting Between D°M′S″ and Decimal Degree Measure. Then press [ENTER]. We find that sin 84°10′39″ $\approx$ 0.9948.

FINDING ANGLES

The inverse trigonometric function keys provide a quick way to find an angle given a trigonometric function value for that angle.

Example 8, page 123: Find the acute angle, to the nearest tenth of a degree, whose sine value is approximately 0.20113.

With the grapher set in Degree mode press [2nd] [SIN^{-1}] [·] 2 0 1 1 3 [ENTER]. (SIN^{-1} is the second operation associated with the [SIN] key.) We find that the desired acute angle is approximately 11.6°.

Exercise 63, page 126: Find the acute angle, to the nearest tenth of a degree, whose cotangent value is 2.127.

Angles whose secant, cosecant, or cotangent values are known can be found using the reciprocals of the cosine, sine, and tangent functions, respectively. Since $\cot\theta = \dfrac{1}{\tan\theta} = 2.127$, we have $\tan\theta = \dfrac{1}{2.127}$, or $(2.127)^{-1}$. To find θ press [2nd] [TAN^{-1}] [(] 1 [÷] 2 [·] 1 2 7 [)] [ENTER] or [2nd] [TAN^{-1}] 2 [·] 1 2 7 [2nd] [x^{-1}] [ENTER]. (TAN^{-1} is the second operation associated with the [TAN] key.) Note that the parentheses are necessary in the first set of keystrokes. Without parentheses we would be finding the angle whose tangent is 1 and then dividing that angle by 2.127. We find that $\theta \approx$ 25.2°.

```
sin-1 .20113
            11.6030461313
tan-1 (1/2.127)
            25.1803638359
tan-1 2.127-1
            25.1803638359
■
```

CONVERTING BETWEEN DEGREE AND RADIAN MEASURE

We can use the grapher to convert from degree to radian measure and vice versa. The grapher should be set in Radian mode when converting from degree to radian measure and in Degree mode when converting from radian to degree measure.

Example 3, page 157: Convert each of the following to radians.

a) 120° b) −297.25°

a) Set the grapher in Radian mode. Press 1 2 0 [2nd] [MATH] [F3] [F1] [ENTER] to enter 120°. The grapher returns a decimal approximation of the radian measure. We see that 120° ≈ 2.09 radians.

b) With the grapher set in Radian mode and with the ANGLE menu displayed, press [(−)] 2 9 7 [·] 2 5 [F1] [ENTER]. We see that −297.25° ≈ −5.19 radians.

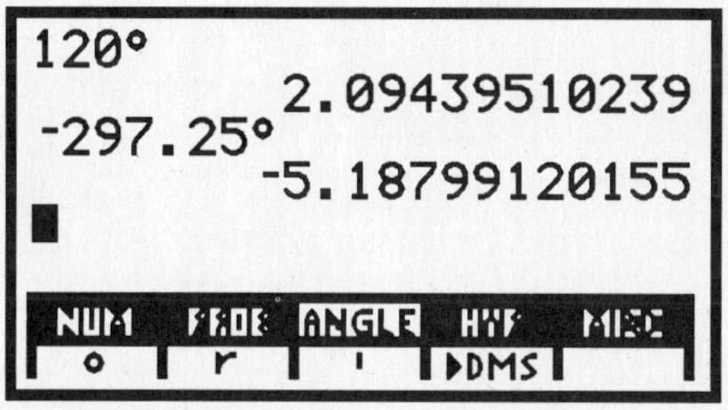

Example 4, page 157: Convert each of the following to degrees.

a) $\frac{3\pi}{4}$ radians b) 8.5 radians

a) Set the grapher in Degree mode. Then press [(] 3 [2nd] [π] [÷] 4 [)] [2nd] [MATH] [F3] [F2] [ENTER] to enter $\frac{3\pi}{4}$ radians. (π is the second operation associated with the [∧] key). The grapher returns 135, so $\frac{3\pi}{4}$ radians = 135°. Note that the parentheses are necessary in order to enter the entire expression in radian measure. Without the parentheses, the grapher reads only the denominator, 4, in radian measure and an incorrect result occurs.

b) With the grapher set in Degree mode and with the ANGLE menu displayed, press 8 [·] 5 [F2] [ENTER]. The grapher returns 487.014125861, so 8.5 radians ≈ 487.01°.

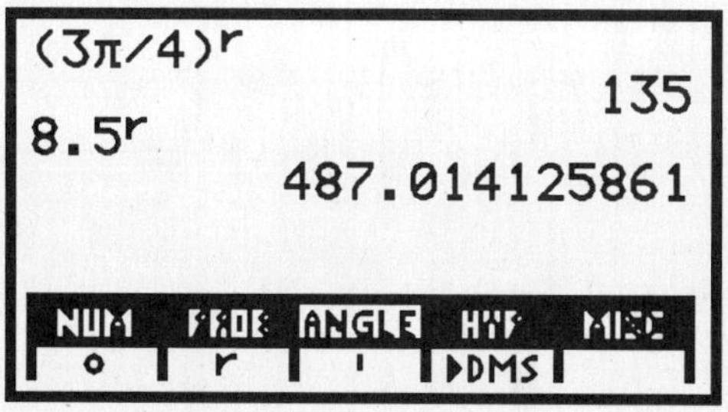

Chapter 3
Trigonometric Identities, Inverse Functions, and Equations

FINDING INVERSE FUNCTION VALUES

We can use a grapher to find inverse function values in both radians and degrees.

Example 2 (a), (e), page 235: Approximate $\cos^{-1}(-0.2689)$ and $\csc^{-1} 8.205$ in both radians and degrees.

To find inverse function values in radians, first set the grapher in Radian mode. Then, to approximate $\cos^{-1}(-0.2689)$, press [2nd] [COS^{-1}] [(−)] . 2 6 8 9 [ENTER]. The grapher returns 1.84304711148, so $\cos^{-1}(-0.2689) \approx 1.8430$ radians.

To find $\csc^{-1} 8.205$, recall the identity $\csc\theta = \dfrac{1}{\sin\theta}$. Then $\csc^{-1} 8.205 = \sin^{-1}\left(\dfrac{1}{8.205}\right)$. Press [2nd] [SIN^{-1}] [(] 1 [÷] 8 . 2 0 5 [)] [ENTER] or [2nd] [SIN^{-1}] 8 . 2 0 5 [2nd] [x^{-1}] [ENTER]. The readout is .122180665346, so $\csc^{-1} 8.205 \approx 0.1222$ radians.

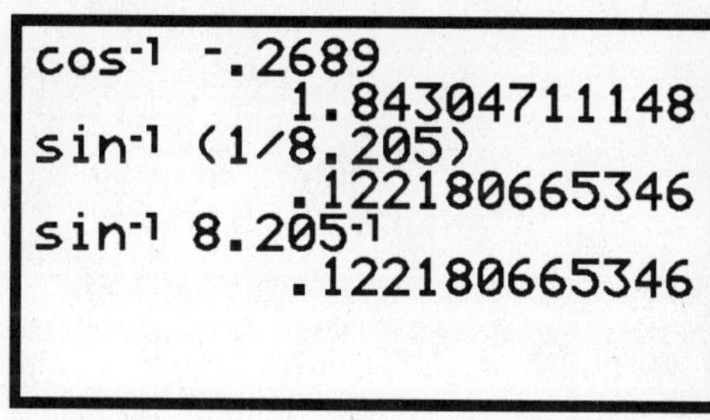

To find inverse function values in degrees, set the grapher in degree mode. Then use the keystrokes above to find that $\cos^{-1}(-0.2689) \approx 105.6°$ and $\csc^{-1} 8.205 \approx 7.0°$. We also use reciprocal relationships to find function values for arcsecant and arccotangent.

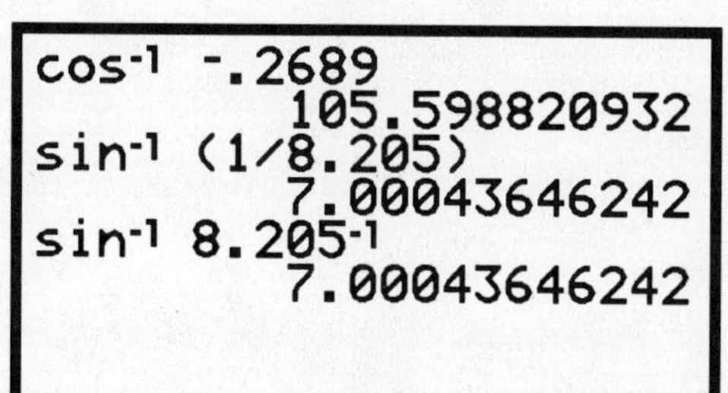

Chapter 4
Applications of Trigonometry

OPERATIONS WITH COMPLEX NUMBERS

Operations with complex numbers can be performed on the TI-85. First check to see that the grapher is set in rectangular complex mode. If it is not, press [2nd] [MODE], position the cursor over RectC, and press [ENTER].

Example 4, page 281:

(a) Add: $(8+6i)+(3+2i)$.

A complex number $a+bi$ is entered on the TI-85 as the ordered pair (a,b). Thus, to find this sum press [(] 8 [,] 6 [)] [+] [(] 3 [,] 2 [)] [ENTER]. The grapher returns (11,8) which corresponds to the number $11+8i$.

(b) Subtract: $(4+5i)-(6-3i)$.

Press [(] 4 [,] 5 [)] [−] [(] 6 [,] [(−)] 3 [)] [ENTER]. The grapher returns $(-2,8)$ which corresponds to the number $-2+8i$.

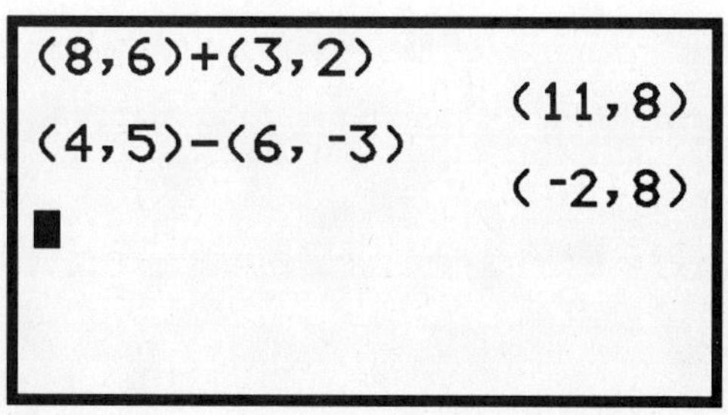

Example 5 (a), (b), (d), (e), page 282:

(a) Multiply: $\sqrt{-16}\cdot\sqrt{-25}$.

Press [2nd] [√] [(] [(−)] 1 6 [)] [2nd] [√] [(] [(−)] 2 5 [)] [ENTER]. The grapher returns $(-20,0)$ which indicates that $\sqrt{-16}\cdot\sqrt{-25}=-20+0i$, or -20.

(b) Multiply: $\sqrt{-5}\cdot\sqrt{-7}$.

Press [2nd] [√] [(] [(−)] 5 [)] [2nd] [√] [(] [(−)] 7 [)] [ENTER]. This operation produces a decimal approximation of the product, $(-5.9160797831,0)$ That is, $\sqrt{-5}\cdot\sqrt{-7}\approx-5.9160797831$. The exact value of the product is $-\sqrt{35}$.

```
√(-16)√(-25)
                 (-20,0)
√(-5)√(-7)
     (-5.9160797831,0)
```

(d) Multiply: $(1+2i)(1+3i)$.

Press [(] [1] [,] [2] [)] [(] [1] [,] [3] [)] [ENTER]. The result is $(-5,5)$, or $-5+5i$.

(e) Multiply: $(3-7i)^2$.

Press [(] [3] [,] [(−)] [7] [)] [x^2] [ENTER]. The result is $(-40,-42)$, or $-40-42i$.

```
(1,2)(1,3)
                  (-5,5)
(3,-7)²
              (-40,-42)
```

Complex numbers can also be divided in RectC mode. The real and imaginary parts of the quotient will be expressed in decimal form. For example, the quotient $(2-5i)/(1-6i)$ in Example 9 on page 284 of the textbook would be expressed as (.864864864865, .189189189189).

Chapter 5
Analytic Geometry Topics

Many conic sections are represented by equations that are not functions. Consequently, these equations must be entered on the TI-85 as two equations, each of which is a function.

GRAPHING PARABOLAS

To graph a parabola of the form $y^2 = 4px$ or $(y-k)^2 = 4p(x-h)$, we must first solve the equation for y.

Example 4, page 328: Graph the parabola $y^2 - 2y - 8x - 31 = 0$.

In the text we used the quadratic formula to solve the equation for y:

$$y = \frac{2 \pm \sqrt{32x+128}}{2}.$$

One way to produce the graph of the parabola is to enter $y_1 = \dfrac{2+\sqrt{32x+128}}{2}$ and $y_2 = \dfrac{2-\sqrt{32x+128}}{2}$, select a window, and press [GRAPH] to see the graph. Here we use $[-12, 12, -7, 7]$ since a square window gives the best representation of the shape of the parabola. The first equation produces the top half of the parabola and the second equation produces the lower half.

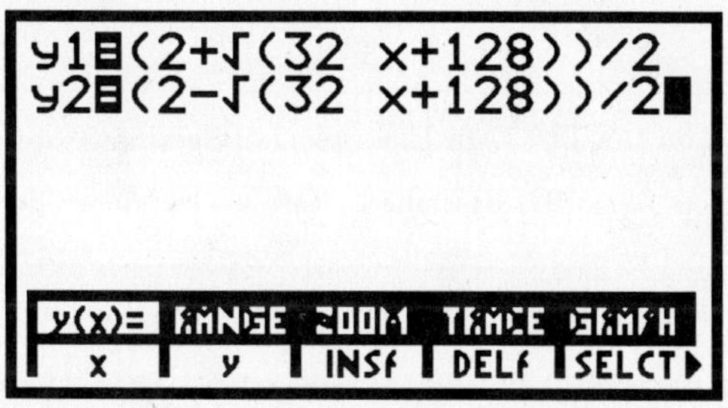

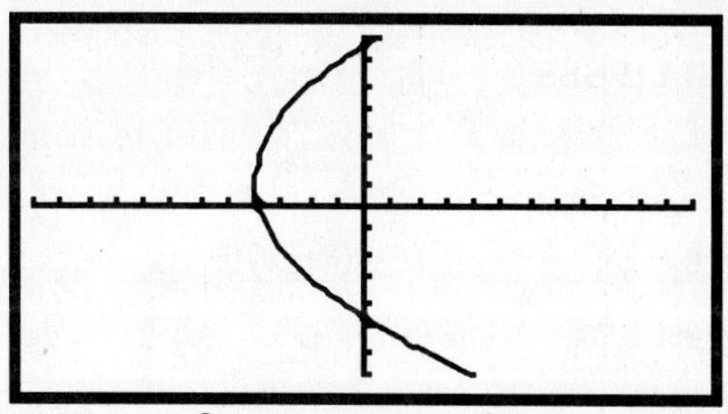

We can also enter $y_1 = \sqrt{32x+128}$. Then use $y1$ to enter $y_2 = \dfrac{2+y_1}{2}$ and $y_3 = \dfrac{2-y_1}{2}$. For example, to enter $y_2 = \dfrac{2+y_1}{2}$ position the cursor beside "y2 =" and press [(] 2 [+] [F2] 1 [)] [÷] 2. Enter $y_3 = \dfrac{2-y_1}{2}$ in a similar manner. Finally, deselect y_1 by moving the cursor to the right of the equals sign following y1 and pressing [F5] to select SELCT. The top half of the graph is produced by y_2 and the lower half by y_3. The expression for y_1 was entered to avoid entering the square root more than once. By deselecting y_1 we prevent the graph of y_1 from appearing on the screen with the graph of the parabola.

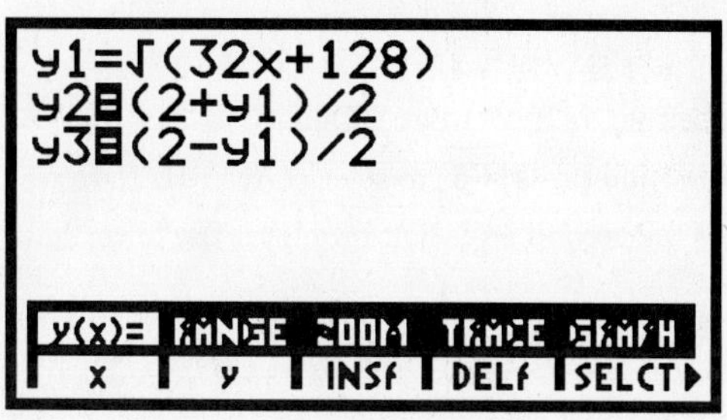

We could also use the standard equation of the parabola found in the text:

$$(y-1)^2 = 8(x+4).$$

Solve this equation for y.

$$y-1 = \pm\sqrt{8(x+4)}$$
$$y = 1 \pm \sqrt{8(x+4)}$$

Then enter $y_1 = 1+\sqrt{8(x+4)}$ and $y_2 = 1-\sqrt{8(x+4)}$, or enter $y_1 = \sqrt{8(x+4)}$, $y_2 = 1+y_1$, and $y_3 = 1-y_1$, and deselect y_1 as above.

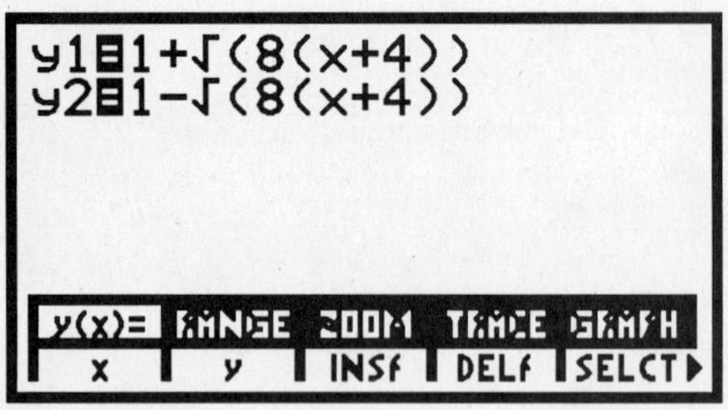

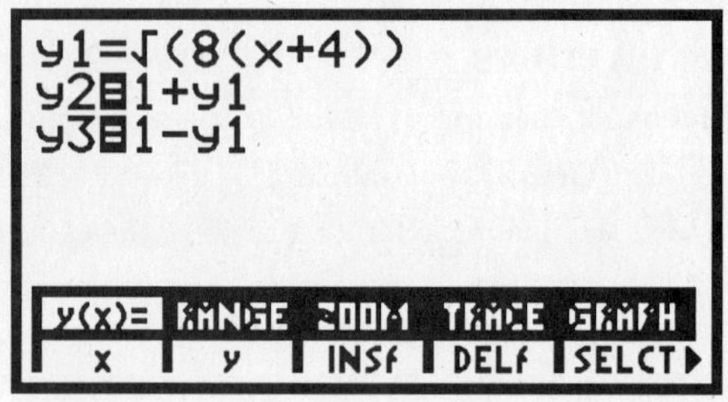

GRAPHING CIRCLES

See Appendix A.2 of this manual for a discussion of graphing circles on a grapher.

GRAPHING ELLIPSES

The equation of an ellipse must be solved for y before it can be entered on the TI-85. In Example 2 of Section 8.2 of the text the procedure for graphing an ellipse of the form $\frac{x^2}{a^2}+\frac{y^2}{b^2}=1$ or $\frac{x^2}{b^2}+\frac{y^2}{a^2}=1$ is described. Here we consider ellipses of the form $\frac{(x-h)^2}{a^2}+\frac{(y-k)^2}{b^2}=1$ or $\frac{(x-h)^2}{b^2}+\frac{(y-k)^2}{a^2}=1$

Example 4, page 338: Graph the ellipse $4x^2+y^2+24x-2y+21=0$.

Completing the square in the text, we found that the equation can be written as

$$\frac{(x+3)^2}{4}+\frac{(y-1)^2}{16}=1.$$

Solve this equation for y.

$$\frac{(x+3)^2}{4}+\frac{(y-1)^2}{16}=1$$
$$\frac{(y-1)^2}{16}=1-\frac{(x+3)^2}{4}$$
$$(y-1)^2 = 16-4(x+3)^2 \qquad \text{Multiplying by 16}$$
$$y-1 = \pm\sqrt{16-4(x+3)^2}$$
$$y = 1\pm\sqrt{16-4(x+3)^2}$$

Now we can produce the graph in either of two ways. One is to enter $y_1 = 1+\sqrt{16-4(x+3)^2}$ and $y_2 = 1-\sqrt{16-4(x+3)^2}$, select a square window, and press F5 to select GRAPH. Here we use $[-12, 12, -7, 7]$. The first equation produces the top half of the ellipse and the second equation produces the lower half.

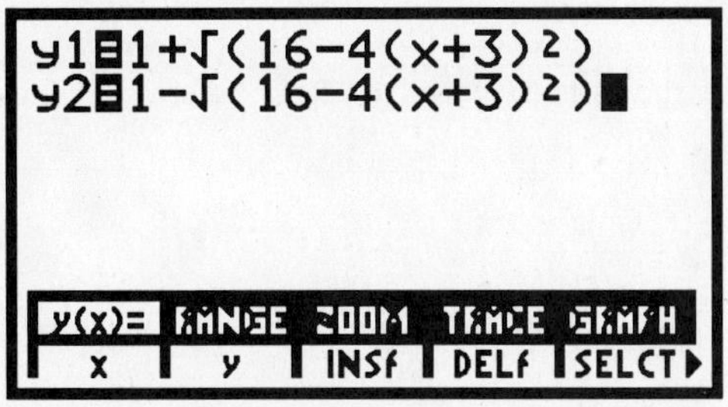

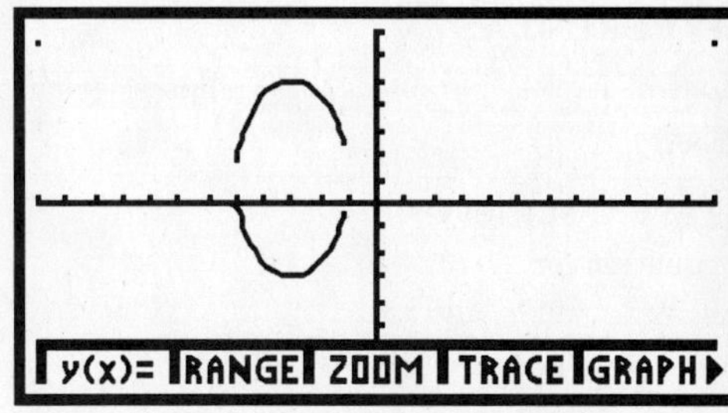

We can also enter $y_1 = \sqrt{16 - 4(x+3)^2}$ and use F2 to select y as described above in Graphing Parabolas to enter $y_2 = 1 + y_1$ and $y_3 = 1 - y_1$. Then deselect y_1, select a square window, and finally select GRAPH. As with parabolas, y_1 is used to eliminate the need to enter the square root more than once. Deselecting it prevents the graph of y_1 from appearing on the screen with the graph of the ellipse. The top half of the graph is produced by y_2 and the lower half by y_3.

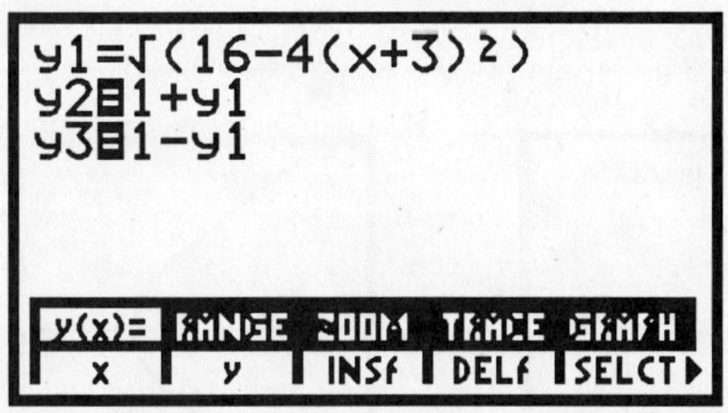

We could also begin by using the quadratic formula to solve the original equation for y.

$$4x^2 + y^2 + 24x - 2y + 21 = 0$$

$$y^2 - 2y + (4x^2 + 24x + 21) = 0$$

$$y = \frac{-(-2) \pm \sqrt{(-2)^2 - 4 \cdot 1 \cdot (4x^2 + 24x + 21)}}{2 \cdot 1}$$

$$y = \frac{2 \pm \sqrt{4 - 16x^2 - 96x - 84}}{2}$$

$$y = \frac{2 \pm \sqrt{-16x^2 - 96x - 80}}{2}$$

Then enter $y_1 = \dfrac{2 + \sqrt{-16x^2 - 96x - 80}}{2}$ and $y_2 = \dfrac{2 - \sqrt{-16x^2 - 96x - 80}}{2}$, or enter $y_1 = \sqrt{-16x^2 - 96x - 80}$, $y_2 = \dfrac{2 + y_1}{2}$, and $y_3 = \dfrac{2 - y_1}{2}$, and deselect y_1.

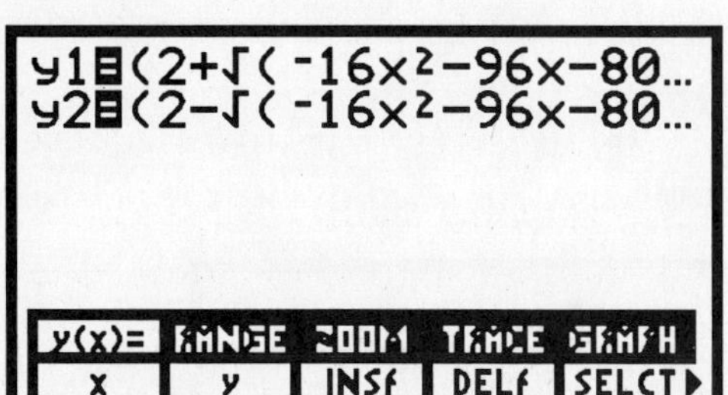

Select a square window and press F5 to display the graph.

GRAPHING HYPERBOLAS

As with equations of circles, parabolas, and ellipses, equations of hyperbolas must be solved for y before they can be entered on the TI-85.

Example 2, page 344: Graph the hyperbola $9x^2 - 16y^2 = 144$.

First solve the equation for y.

$$\begin{aligned} 9x^2 - 16y^2 &= 144 \\ -16y^2 &= -9x^2 + 144 \\ y^2 &= \frac{-9x^2 + 144}{-16} \\ y &= \pm\sqrt{\frac{-9x^2 + 144}{-16}}, \text{ or } \pm\sqrt{\frac{9x^2 - 144}{16}} \end{aligned}$$

It is not necessary to simplify further.

Now enter $y_1 = \sqrt{\dfrac{9x^2 - 144}{16}}$ and either $y_2 = -\sqrt{\dfrac{9x^2 - 144}{16}}$ or $y_2 = -y_1$, select a square window, and press F5 to select GRAPH. Here we use $[-12, 12, -7, 7]$. The top half of the graph is produced by y_1 and the lower half by y_2.

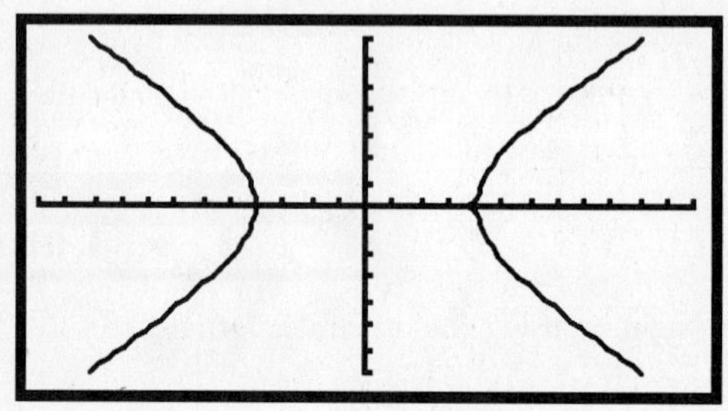

Example 3, page 347: Graph the hyperbola $4y^2 - x^2 + 24y + 4x + 28 = 0$.

In the text we completed the square to get the standard form of the equation. Now solve the equation for y.

$$\begin{aligned} \frac{(y+3)^2}{1} - \frac{(x-2)^2}{4} &= 1 \\ (y+3)^2 &= \frac{(x-2)^2}{4} + 1 \\ y + 3 &= \pm\sqrt{\frac{(x-2)^2}{4} + 1} \\ y &= -3 \pm \sqrt{\frac{(x-2)^2}{4} + 1} \end{aligned}$$

As with ellipses, the graph can be produced in either of two ways. One way is to enter $y_1 = -3 + \sqrt{\dfrac{(x-2)^2}{4} + 1}$ and $y_2 = -3 - \sqrt{\dfrac{(x-2)^2}{4} + 1}$, select a square window, and press F5 to select GRAPH. Here we use $[-12, 12, -7, 7]$. The first equation produces the top half of the hyperbola and the second the lower half.

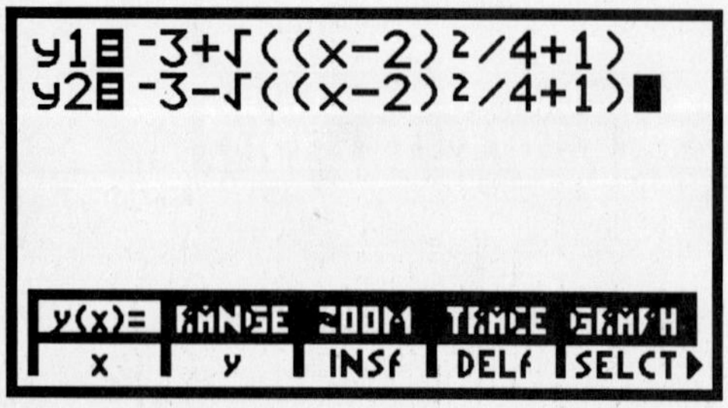

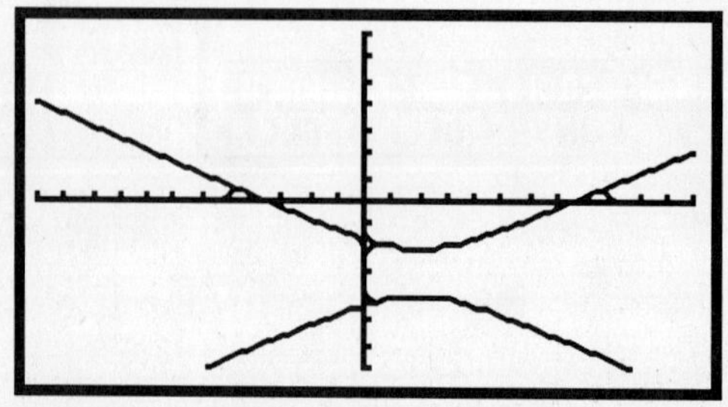

We can also enter $y_1 = \sqrt{\dfrac{(x-2)^2}{4} + 1}$, $y_2 = -3 + y_1$, and $y_3 = -3 - y_1$. Then deselect y_1, select a square window, and press F5 to select GRAPH. As with parabolas and ellipses, y_1 is used to eliminate the need to enter the square root more than once. Deselecting it prevents the graph of y_1 from appearing on the screen with the graph of the hyperbola. The top half of the graph is produced by y_2 and the lower half by y_3.

CONVERTING FROM RECTANGULAR TO POLAR COORDINATES

The grapher can be used to convert from rectangular to polar coordinates, expressing the result using either degrees or radians. The grapher will supply a positive value for r and an angle in the interval $(-180°, 180°]$, or $(-\pi, \pi]$.

Example 2 (a), page 370: Convert (3,3) to polar coordinates.

Set the grapher in Degree mode and PolarC mode. Then press (3 , 3) ENTER. The readout is (4.24264068712∠45), so $r \approx 4.2426$ (a decimal approximation for $3\sqrt{2}$) and $\theta = 45°$. Thus, polar notation for (3,3) is (4.2426,45°).

Set the grapher in Radian mode and PolarC mode to find θ in radians. Repeat the keystrokes for finding (3,3) above to find that r is the same and $\theta \approx 0.7854$ (a decimal approximation for $\pi/4$). Thus polar notation for (3,3) is (4.2426,0.7854).

CONVERTING FROM POLAR TO RECTANGULAR COORDINATES

The grapher can also be used to convert from polar to rectangular coordinates.

Example 3, page 371: Convert each of the following to rectangular coordinates.

(a) $(10, \pi/3)$ (b) $(-5, 135°)$

(a) Set the grapher in RectC mode and, since the angle is given in radians, set it in Radian mode also. Press [( ] 1 0 [2nd] [∠] [2nd] [π] [÷] 3 [ )] [ENTER]. (∠ is the second operation associated with the [,] key.) The readout is (5,8.66025403784), where 8.66025403784 is a decimal approximation of $5\sqrt{3}$. Thus, rectangular notation for $(10, \pi/3)$ is (5,8.6603).

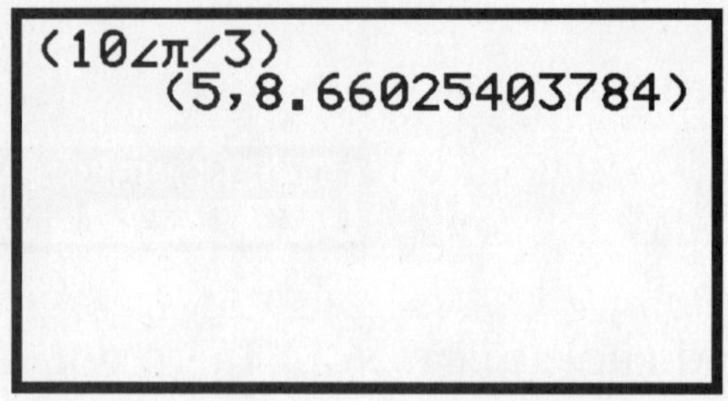

(b) Set the grapher in RectC mode and, since the angle is given in degrees, set it in Degree mode also. To find the x-coordinate of rectangular notation, press [( ] [(−)] 5 [2nd] [∠] 1 3 5 [ )] [ENTER]. The readout is (3.53553390593, −3.53553390593), where 3.53553390593 is a decimal approximation of $\frac{5\sqrt{2}}{2}$. Thus, rectangular notation for $(-5, 135°)$ is $(3.5355, -3.5355)$.

GRAPHING POLAR EQUATIONS

Polar equations can be graphed in either Radian mode or Degree mode. The equation must be written in the form $r = f(\theta)$ and the grapher must be set in polar (Pol) mode. Typically we begin with a range of $[0, 2\pi]$ or [0°,360°], but it might be necessary to increase the range to ensure that sufficient points are plotted to display the entire graph.

Example 6, page 373: Graph: $r = 1 - \sin\theta$.

First set the grapher in Pol mode and in Radian mode.

The equation is given in $r = f(\theta)$ form. Press [GRAPH] [F1] to enter it on the "r(θ) =" screen. Clear any existing entries and, with the cursor beside "r1 =" press 1 [−] [SIN] [F1]. The keystroke [F1] selects the variable θ. Now press [2nd] [M2] to select RANGE and enter the following settings:

θMin = 0	(Smallest value of θ to be evaluated)
θMax = 2π	(Largest value of θ to be evaluated)
θStep = $\pi/24$	(Increment in θ values)
xMin = −6	
xMax = 6	
xScl = 1	
yMin = −3.5	
yMax = 3.5	
yScl = 1	

With these settings the grapher evaluates the function from $\theta = 0$ to $\theta = 2\pi$ in increments of $\pi/24$ and displays the graph in the square window $[-6, 6, -3.5, 3.5]$. Values entered in terms of π appear on the screen as decimal approximations. Press [F5] to display the graph.

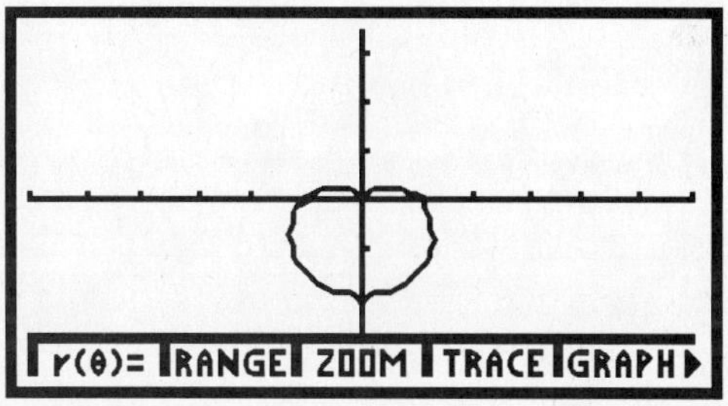

Although we have used a square window, the software used to produce the graph above causes some distortion. The graph produced on your grapher will not be distorted.

The curve can be traced with either rectangular or polar coordinates being displayed. The value of θ is also displayed when rectangular coordinates are selected. The choice of coordinates is made on the GRAPH FORMT screen. Press [MORE] [F3] to display this screen. Then position the blinking cursor over RectGC to select rectangular coordinates or over PolarGC to select polar coordinates.

Chapter 6
Exponential and Logarithmic Functions

EVALUATING e^x, Log x, and Ln x

Use the grapher's scientific keys to evaluate e^x, $\log x$, and $\ln x$ for specific values of x.

Example 6 (a), (b), page 407: Find the value of e^3 and $e^{-0.23}$. Round to four decimal places.

To find e^3 press [2nd] [e^x] 3 [ENTER]. (e^x is the second operation associated with the [LN] key.) The grapher returns 20.0855369232. Thus, $e^3 \approx 20.0855$. To find $e^{-0.23}$ press [2nd] [e^x] [(−)] [.] 2 3 [ENTER]. The grapher returns .794533602503, so $e^{-0.23} \approx 0.7945$.

Example 4 (a), (b), (c), page 416: Find the values of log 645,778, log 0.0000239, and log (−3). Round to four decimal places.

To find log 645,778 press [LOG] 6 4 5 7 7 8 [ENTER] and read 5.81008324563. Thus, $\log\ 645,778 \approx 5.8101$. To find log 0.0000239 press [LOG] [.] 0 0 0 0 2 3 9 [ENTER]. The grapher returns −4.62160209905, so $\log 0.0000239 \approx -4.6216$. When we press [LOG] [(−)] 3 [ENTER] the result is a non-real complex number. This indicates that −3 is not in the domain of the function $\log x$.

Example 5 (a), (b), (c), page 416: Find the values of ln 645,778, ln 0.0000239, and ln (−5). Round to four decimal places.

To find ln 645,778 and ln 0.0000239 repeat the keystrokes used above to find log 645,778 and log 0.0000239 but press [LN] rather than [LOG]. We find that ln 645,778 ≈ 13.3782 and $\ln 0.0000239 \approx -10.6416$. When we press [LN] [(−)] 5 [ENTER] the result is a non-real complex number, indicating that −5 is not in the domain of the function $\ln x$.

USING THE CHANGE OF BASE FORMULA

To find a logarithm with a base other than 10 or e we use the change-of-base formula, $\log_b M = \dfrac{\log_a M}{\log_a b}$, where a and b are any logarithmic bases and M is any positive number.

Example 6, page 417: Find $\log_5 8$ using common logarithms.

We let $a = 10$, $b = 5$, and $M = 8$ and substitute in the change-of-base formula. Press [LOG] 8 [÷] [LOG] 5 [ENTER]. The result is about 1.2920.

Example 7, page 417: Find $\log_4 31$ using natural logarithms.

We let $a = e$, $b = 4$, and $M = 31$ and substitute in the change-of-base formula. Press [LN] 3 1 [÷] [LN] 4 [ENTER]. The result is about 2.4771.

Example 9, page 418: Graph $y = \log_5 x$.

To use a grapher we must first change the base to e or 10. Here we use e. Let $a = e$, $b = 5$, and $M = x$ and substitute in the change-of-base formula. Enter $y_1 = \dfrac{\ln x}{\ln 5}$ on the "y(x) =" screen, select a window, and select GRAPH from the menus.

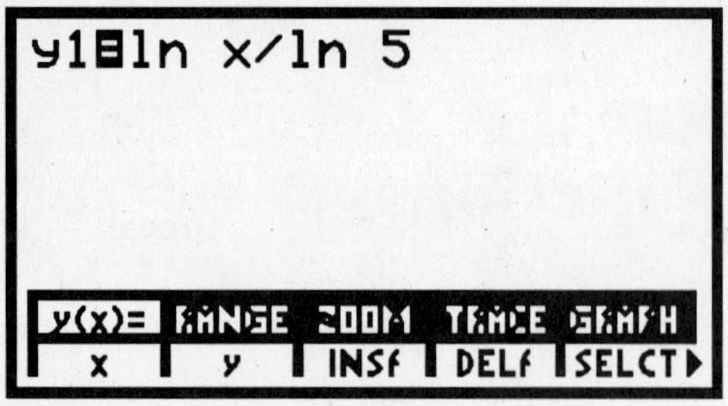

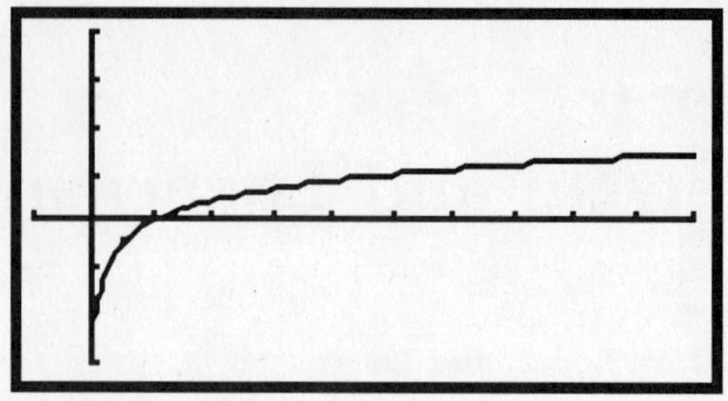

EXPONENTIAL, LOGARITHMIC, AND POWER REGRESSION

Exponential, logarithmic, and power functions can be fit to data. The operations of entering data, making scatterplots, and graphing and evaluating these functions are the same as for linear regression functions. So are the procedures for copying a regression equation to the "y(x) =" screen, graphing it, and using it to find function values. See Appendix A.3 of this manual (page 111) for a discussion of these procedures.

Example 6 (a), page 447: Fit an exponential equation to the given data on cellular phones.

Enter the data in xStat and yStat as described on page 111 of this manual. Then press [2nd] [M1] [ENTER] [ENTER] [F4] to select EXPR. The values of a and b for the exponential function $y = ab^x$ are displayed along with the coefficient of correlation and the number of data points n.

Exercise 26 (a), page 453: Fit a logarithmic function to the given data on forgetting.

After entering the data in xStat and yStat as described on page 111 of this manual, press [2nd] [M1] [ENTER] [ENTER] [F3] to select LNR. The values of a and b for the logarithmic function $y = a + b \ln x$ are displayed along with the coefficient of correlation and the number of data points n.

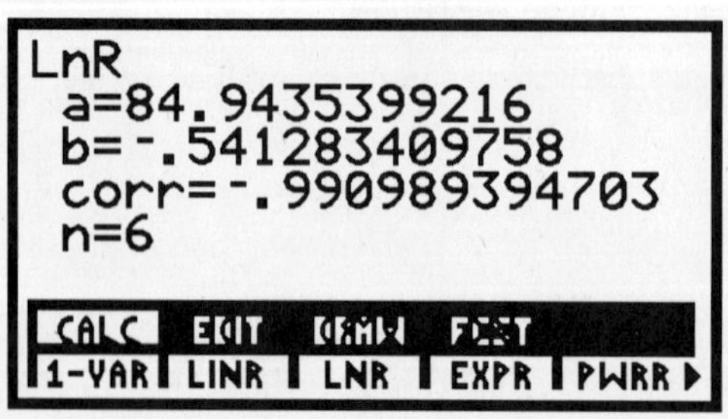

Example 7 (a), page 449: Fit a power function to the given data on cholesterol level and risk of heart attack.

Enter the data in xStat and yStat as described on page 111 of this manual. Then press [2nd] [M1] [ENTER] [ENTER]

F5 to select PWRR. The values of a and b for the power function $y = ax^b$ are displayed along with the coefficient of correlation and the number of data points n.

Appendix A.1
Solving Equations

CHECKING SOLUTIONS USING A TABLE

The program "ATABLE" on page 117 of this manual can be used to give the TI-85 powerful tables which can be used to check solutions of equations.

Appendix A.2
Distance, Midpoints, and Circles

GRAPHING CIRCLES

Although the standard form of the equation of a circle cannot be entered in the "y(x) =" screen, the TI-85 can be used to graph circles. A square viewing window is required to give an accurate graph using either of the methods discussed below.

Method 1: We begin by solving the equation for y.

Example 5, page 474: Graph $x^2 + y^2 = 16$.

Solve the equation for y as shown on page 474 of the text, enter $y_1 = \sqrt{16 - x^2}$ and $y_2 = -\sqrt{16 - x^2}$, select a square window, and press [GRAPH]. Note that we can enter $y_2 = -y_1$ instead of $y_2 = -\sqrt{16 - x^2}$. The graph is shown on page 138 of the text.

Method 2: If the center and radius of a circle are known, the circle can be graphed by typing an instruction on the home screen.

Exercise 37, page 475: Graph $(x - 1)^2 + (y - 5)^2 = 36$.

The center of this circle is (1,5) and its radius is 6. To graph it using the DRAW feature first press [GRAPH] [F1] to select y(x) = and clear all previously entered equations. Then select a square window. We will use $[-12, 12, -2, 12]$. Press [2nd] [QUIT] to go to the home screen. Now press [ALPHA] [ALPHA] and type CIRCL. Press [ALPHA] [(]. Enter the coordinates of the center and the radius, separating the entries by commas, and close the parentheses: 1 [,] 5 [,] 6 [)] [ENTER].

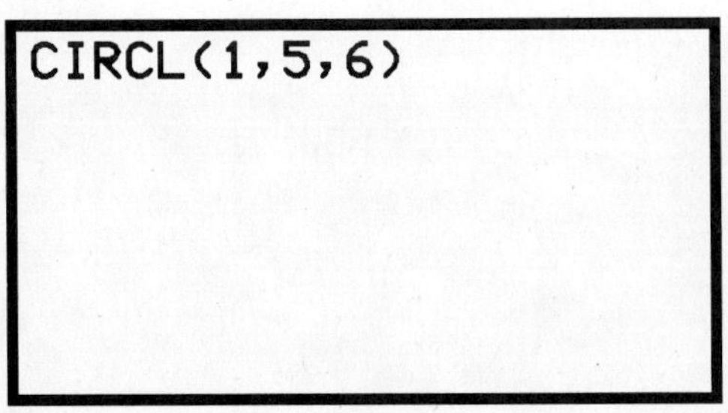

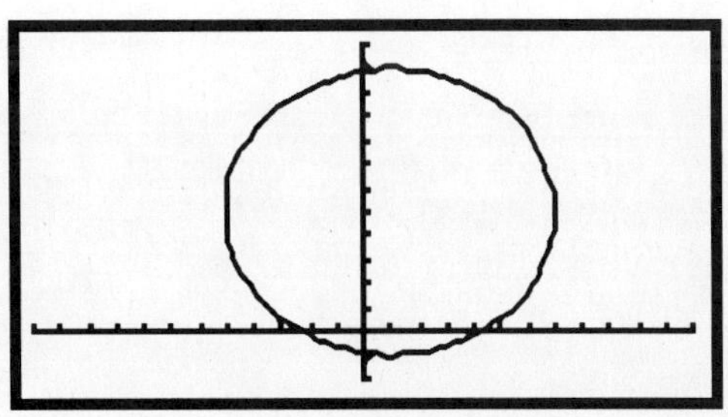

As before, the software used to produce the the graph above causes distortion. Nevertheless, when the circle is graphed on a grapher in a square window, there is no distortion.

A circle graphed from the home screen can be cleared from the graph screen by pressing [MORE] [F2] to access the DRAW menu. Then press [MORE] [F5] to select the CLDRW (clear drawing) operation. The graph will also be cleared when another function is subsequently entered on the "y(x) =" screen and graphed.

A circle can also graphed using the DRAW feature, but this method does not yield as accurate a graph as the other two methods.

Note that the equation of a circle that is not centered at the origin, like the one in Example 37 above, is generally not easily solved for y, making Method 2 preferable for such equations. In fact, since there is no need to solve for y using this method, it provides a quick procedure for graphing the equation of any circle in standard form.

Appendix A.3
Data Analysis, Curve Fitting, and Regression

LINEAR REGRESSION AND SCATTERPLOTS

The linear regression operation of the grapher enables us to fit a linear function to a set of data.

Example 1, page 478: Fit a regression line to the data on life expectancy of women on page 477 of the text. Then predict the life expectancy of women in 2000.

First we enter the data from the table on the STAT list editor screen. To clear any existing lists press [STAT] [F2] [ENTER] [ENTER] [F5] to select CLRxy.

To enter the first x-value press 0 [ENTER]; type the first y value to the right of "y_1 =" and press [ENTER]. Press 1 [ENTER] to enter the second x-value and then type the second y value to the right of "y_2 =." Press [ENTER] and continue in this manner until all pairs of data are entered. However, do not press [ENTER] after the last y-value.

Press [2nd] [M1] to select CALC. Then press [ENTER] [ENTER] [F2] to select LINR and display the coefficients a and b of the regression line $y = a + bx$, the coefficient of linear correlation (denoted "corr"), and the number of data points, n.

To see a scatterplot of the data press [2nd] [M3] to select DRAW and then [F2] to select SCAT. To draw the regression line on the same screen as the scatterplot press [F4] to select DRREG.

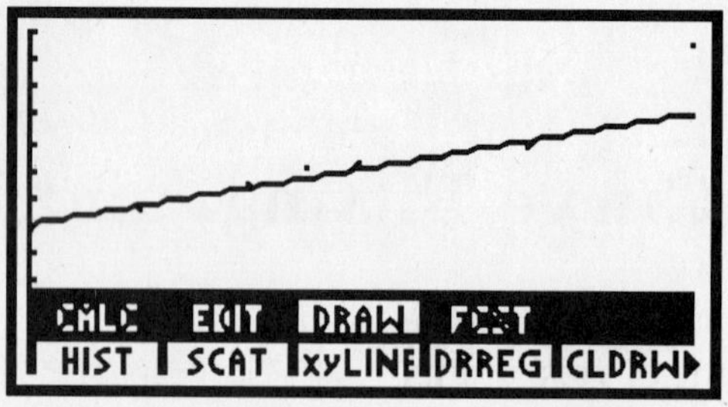

To enter the regression equation as y_1 in the "y(x) =" screen, with the CALC menu and the LinR screen still displayed, press MORE F4 to select STREG (store regression). Then type $y1$ to the right of the blinking cursor beside "Name =" by pressing 2nd alpha y 2nd alpha 1 ENTER. (alpha is the second operation associated with the ALPHA key. It is used when we want to type lowercase letters.)

To predict the life expectancy of women in 2000 find the value of y_1 when $x = 10$ using one of the methods described in "Finding Function Values" on page 81 of this manual.

QUADRATIC, CUBIC, AND QUARTIC REGRESSION

Second degree, third degree, and fourth degree polynomial functions can be fit to data using the quadratic, cubic, and quartic regression operations, respectively.

The operations of entering data, making scatterplots, and graphing and evaluating quadratic, cubic, and quartic regression functions are the same as for linear regression functions.

Example 2, page 481: For the data on hours of sleep versus death rate on page 481 of the text, make a scatterplot for the data and determine which, if any, of the functions above fits the data. Then find the regression equation, graph the equation, and use it to make predictions.

Enter the data in xStat and yStat and make a scatterplot as described above for linear regression. The scatterplot is shown in the text. It appears that a quadratic function fits the data. To find it press 2nd M1 ENTER ENTER MORE F1 to select quadratic regression, denoted as P2REG. The coefficients of the quadratic equation $y = ax^2 + bx + c$ are displayed in parentheses. The number of data points n is also displayed. Note that at least three data points are required for quadratic regression.

The regression equation can be copied to the "y(x) =" screen as described in the discussion of linear regression above.

Then it can be graphed and used to make predictions using one of the methods for finding function values described on page 81.

To fit a cubic or quartic function to a set of data, enter the data in xStat and yStat as described earlier. Then press [2nd] [M1] [ENTER] [ENTER] [MORE]. Select cubic regression by pressing [F2] to select P3REG and quartic regression by pressing [F4] to select P4REG. At least four data points are required for cubic regression and at least five are required for quartic regression.

Programs For The TI-85 Graphics Calculator

PROGRAMS FOR THE TI-85 GRAPHICS CALCULATOR

(Created by Dave Slomer, Winton Woods High School, Cincinnati, OH.)

ATABLE

This program gives the TI-85 powerful TABLES (written by Marco Radzinschi, a freshman at Pikeville High School, Baltimore, MD), shortened and enhanced by Dave Slomer).

INPUT: Enter and/or select all functions to be included in the table and set the RANGE, keeping in mind that the table values will go from xMax to xMin in steps of xScl. Dimensions must be chosen so that all values from xMin to xMax are in the domains of all selected functions.

OUTPUT: Matrix T, viewed best in the MATRIX editor

```
:Func
:int((xMax-xMin)/xScl+1)→K
:dimL eval xMin
:{K,Ans}→dim T
:Ø→K
:For(x,xMin,xMax,xScl)
:  K+1→K
: li▷vc (x+√-1*eval x)→T(K)
:End
```

QUADFORM

This program solves a quadratic equation of form $Ax^2 + Bx + C = 0$.

INPUT: A, B, C

OUTPUT: Real or complex solutions

```
:Prompt A,B,C
:Disp "ROOTS:"
:B²-4A*C→D
:Disp (-B+√D)/(2A)
:Disp (-B-√D)/(2A)
```

TRISOL9∅

This program solves right triangles ABC in which legs L and M are opposite angles A and B, respectively.

INPUT: One side and any other part (except the right angle, C)

OUTPUT: First, the sides; then, the acute angles

```
:Degree
:Menu(1,"HA",P,2,"HL",Q,3,"LA",R,4,"LL",S)
:
:Lbl P
:Prompt H,A
:9∅-A→B
:H*sin (A)→L
:H*sin (B)→M
:Goto X
:
:Lbl Q
:Prompt H,L
:√ (H^2-L^2)→ M
:sin⁻¹ (L/H)→A
:9∅-A→ B
:Goto X
:
:Lbl R
:Prompt L,A
:9∅-A→B
:L/tan (A)→M
:√ (L^2+M^2)→H
:Goto X
:
:Lbl S
:Prompt L,M
:√ (L^2+M^2)→H
:tan⁻¹ (L/M)→ A
:9∅-A→B
:
:Lbl X
:Disp L,M,H,A,B
```

TRISOL

This program solves triangle ABC, in which sides L, M, and N are opposite angles A, B, and C, respectively. It covers the "no solution" case as well as the "two solutions" case.

INPUT: Three parts of a triangle, including at least one side

OUTPUT: First the sides, then the angles. The AAS case is not included, so in this case it is necessary to find the third angle by hand and then use ASA.

```
:Degree
:Disp "ANGLES A,B,C"
:Disp "SIDES L,M,N"
.Menu(1,"SSS",P,2,"SAS",Q,3,"(A)ASA",R,4,"SSA",S)
:
:Lbl P
:Prompt L,M,N
:cos⁻¹ ((M^2+N^2-L^2)/(2*M*N))→A
:cos⁻¹ ((L^2+N^2-M^2)/(2*L*N))→B
:18Ø-A-B→C
:Goto X
:
:Lbl Q
:Prompt L,C,M
:√(L^2+M^2-2*L*M*cos (C))→N
:cos⁻¹ ((M^2+N^2-L^2)/(2*M*N))→A
:18Ø-A-C→B
:Goto X
:
.Lbl R
:Prompt A,N,B
:18Ø-A-B→C
:sin (A)*N/(sin (C))→L
:sin (B)*N/(sin (C))→M
:Goto X
:
:Lbl S
:Prompt L,M,B
:L*sin (B)/M→Z
:If Z>1:Then:Disp "NO SOL":Stop:End
:sin⁻¹ (Z)→A
:18Ø-A-B→C
:sin (C)*M/(sin (B))→N
:Disp L,M,N,A,B,C
:If B≥A:Stop
:Disp "2ND SOL...":Pause
:18Ø-A→A
:18Ø-A-B→C
:sin (C)*M/(sin (B))→N
:
:Lbl X
:Disp L,M,N,A,B,C
```

DEMOIVRE

This program helps visualize Nth roots of a complex number $A + Bi$.

INPUT: A, B, N

OUPUT: Traceable graph of the N points (A, B) corresponding to the roots

```
:Param
:Degree
:Input "REAL PART:",A
:Input "IMAG PART:",B
:Input "INDEX:",N
:√(A^2+B^2)^(1/N)→R
:If A==Ø:Then
:  18Ø-sign (B)*9Ø
:Else
:  tan^-1 (B/A)+18Ø*(A<Ø)
:End
:Ans/N→H
:xtl=R*cos (t)
:ytl=R*sin (t)
:H→tMin
:H+36Ø→tMax
:36Ø/N→tStep
:1.5R→yMax
:-yMax→yMin
:1.7yMax→xMax
:-xMax→xMin
:Trace
```

LOGBASE

This program computes logarithms of any base.

INPUT: Number (X) and base (B)

OUTPUT: $\log_B(X)$

```
:Prompt X
:Input "NEW BASE:",B
:log (X)/log (B)
```

The HP 48G/48GX Graphic Calculator

Preliminaries

The information in this manual applies to both the HP 48G and the HP 48GX graphic calculators. References to the HP 48G should be understood to refer to the HP 48GX also.

Press [ON] to turn on the HP 48G graphic calculator. ([ON] is the key at the bottom left-hand corner of the keypad.) To adjust the contrast, simultaneously press [ON] and [+] (or [−]) to increase (or decrease) the contrast. To turn the grapher off, press [↱] [OFF]. (The right-shift key [↱] is a light green key located near the lower left-hand corner of the keypad. It activates the right-shift operations which are labeled in green above and to the right of their associated primary keys. OFF is the right-shift operation associated with the [ON] key.) The grapher will turn itself off after ten minutes without any activity.

It very important that the user read the HP 48G Series Quick Start Guide before proceeding. Lessons 1 - 23 and 25 and 26 are particularly helpful.

Press [↱] [MODES] to display the MODE settings. (MODES is the right-shift operation associated withe the [CST] key.)

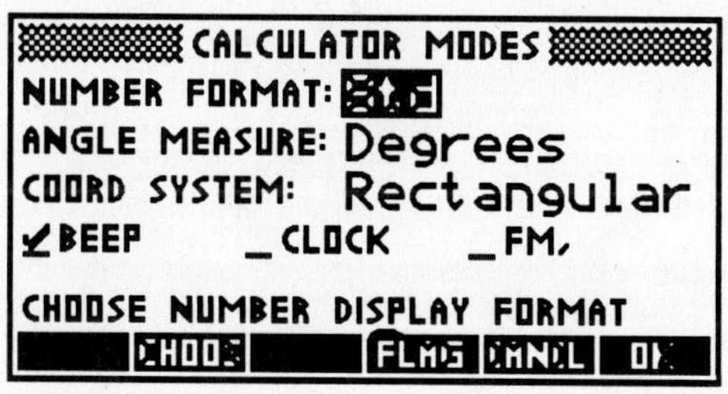

To change a setting use [▽], [△], [▷], or [◁] to move the cursor to that setting. To set the number format, angle measure, or coordinate system, press the sold white key below the screen menu CHOOS, use [▽] or [△] to highlight a setting, and press the white key below the screen item OK or press [ENTER] to select that setting. To set the beeper (which sounds whenever an error occurs), the clock, or FM (fraction mark), use the arrow keys to highlight the setting, and then press the solid white key below √CHK. Finally, press the solid white key below OK. A checkmark can be removed by highlighting it and pressing √CHK followed by the solid white key below OK.

The modes can also be set by using the system flags which can be seen in the Flag Browser which appears in the mode setting screen. Press the white key below FLAG to access the Flag Browser. (Consult the HP 48G User's Guide for more details.)

Throughout this manual we will use double braces to denote the solid white key below a screen menu. For example, {{CHOOS}} denotes the key below CHOOS and {{OK}} denotes the key below OK.

Press {{OK}}, {{CANCL}}, ENTER, or ON to leave the MODES screen.

Chapter G
Introduction to Graphs and Graphers

SETTING THE VIEWING WINDOW

The viewing window is the portion of the coordinate plane that appears on the grapher's screen. H-VIEW specifies the minimum and maximum horizontal values; V-VIEW specifies the minimum and maximum vertical values. H-VIEW and V-VIEW are fields which are found in the Plot setup. The notation [Xmin, Xmax, Ymin, Ymax] is used in the text to represent these window settings or dimensions. For example, [−12, 12, −8, 8] denotes a window that displays the portion of the x-axis from −12 to 12 and the portion of the y-axis from −8 to 8. In addition, the distance between tick marks on the axes is defined by the settings H-TICK and V-TICK. STEP sets the resolution of the plot, which is the distance, measured horizontally, between two consecutive plotted points. A larger step-size will result in faster plots showing less detail, while a smaller step-size will result in slower plots showing greater detail. The default step size, denoted DFLT, for functions is 0.1 units. The PIXELS setting adjacent to the STEP setting determines whether the step-size is interpreted as pixels or units. The default is units and occurs when PIXELS is not checked. The window corresponding to the settings [−20, 30, −10, 20], H-TICK = 5, V-TICK = 2, is shown below.

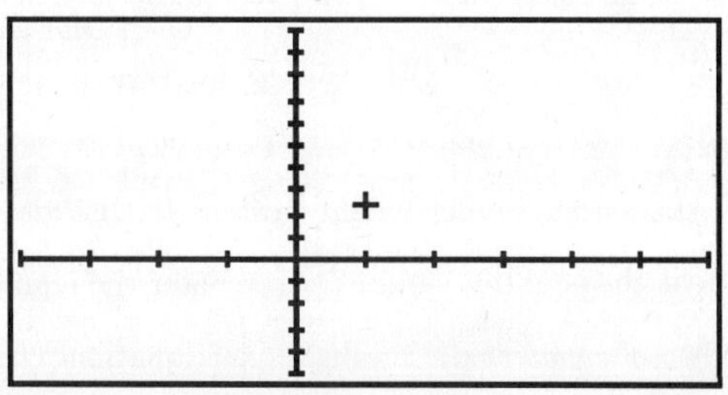

To see the current settings on the grapher, press [↱] [PLOT]. (PLOT is the right-shift operation associated with the [8] key.) The current window settings are displayed on the Plot screen. Additional settings may be accessed from the Plot screen by pressing the menu selection {{OPTS}}.

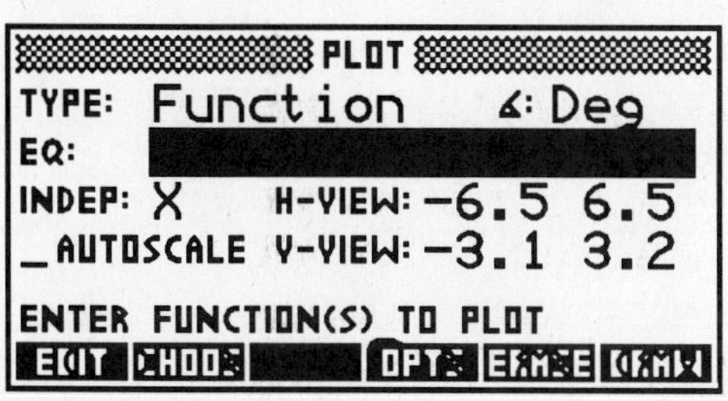

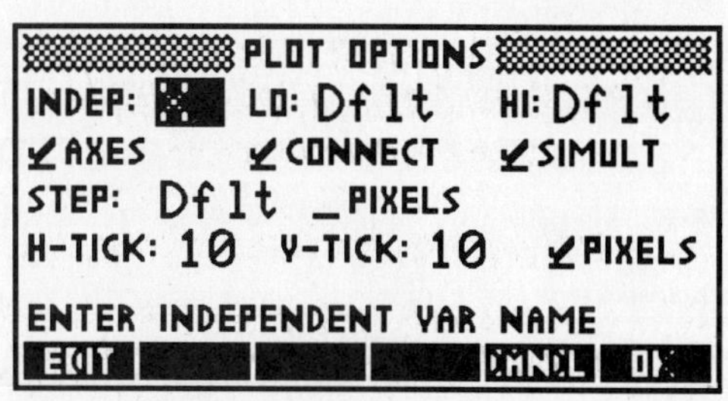

To change a setting use the arrow keys to move the cursor to the setting you wish to change and enter the new value.

For example, to change the settings to [−20, 30, −10, 20], H-TICK = 5, V-TICK = 2, position the cursor over the first setting to the right of H-VIEW and press 2 0 +/− ENTER (or {{OK}}) 3 0 ENTER. Then position the cursor over the first setting to the right of V-VIEW and press 1 0 +/− ENTER 2 0 ENTER. Note that the +/− key on the fifth row of the keypad must be used to enter a negative number rather than the − key in the right-hand column. +/− represents "the opposite of" or "the additive inverse of" whereas − is the subtraction key. Now press {{OPTS}} and use ▽ to highlight the H-TICK setting. Then press 5 ENTER 2 ENTER to reset H-TICK and V-TICK. Also be sure that the PIXELS settings adjacent to both STEP and H- and V-TICK are not checked. If either is, with the cursor positioned over the checkmark beside it, press {{√CHK}} {{OK}}.

To return to the default window setting [−6.5, 6.5, −3.1, 3.2], H-TICK = 10, V-TICK = 10, move the cursor to any setting and press DEL. Then highlight Reset plot on the message window that appears and press ENTER or {{OK}}. All of the window settings will be returned to their default values.

GRAPHING EQUATIONS

Example 8 (a), page 8: (Page numbers refer to pages in the text book.) Graph: $2x + 3y = 18$.

Although an equation must be solved for y before it can be graphed on many graphers, this is not necessary on the HP 48G. First- and second-degree equations in two variables can be entered directly on the HP 48G when the Conic graph-type is chosen.

The equation $2x + 3y = 18$ is a first-degree equation in the two variables x and y, so it does not need to be solved for y before it is entered on the grapher. To produce the graph we begin by pressing ↱ PLOT to access the Plot setup screen. To select the Conic plot-type use △ or ▽ to highlight the TYPE: field, press {{CHOOS}}, use △ or ▽ to highlight Conic, and press ENTER or {{OK}}. Be sure the independent variable INDEP: is set to X. If it is not, highlight INDEP: and press α X ENTER. Now highlight the equation field EQ: and enter the equation $2x + 3y = 18$.

The equation can be entered in either of two ways. One uses the Equation Writer which is opened by pressing ↰ EQUATION. (The left-shift key is a lavender key located on the left-hand side of the keypad. It activates the left-shift operations which are labeled in lavender above and the the left of their associated primary keys. EQUATION is the left-shift operation associated with the ENTER key.) Then press 2 α X + 3 α Y ↰ = 1 8. (= is the left-shift operation associated with the 0 key.) Then press ENTER to enter the equation on the Plot screen.

An entry can be edited in the Equation Writer before it is entered on the Plot screen. If, for instance, you pressed 5 instead of 8, use ◁ to move the cursor to 18 and press {{EDIT}}. This enters the number 18 on the Home screen. Then press ◁ DEL 8 to delete 5 and replace it with 8. Then press ENTER to enter the edited number in the Equation Writer. You can also edit in the Equation Writer using the ⇐ key . If, for example, you had typed 6 instead of 3, press ⇐ repeatedly to delete 6 and all the entries following it. Then reenter the expression beginning with 3. You can also press ON to delete the entire expression and then reenter it. When the editing is complete, press ENTER to enter the equation in the EQ: field of the Plot screen. To edit after entering the expression, highlight it by using the arrow keys to scroll to it and select {{EDIT}}.

The equation can also be entered on the command line of the Plot screen. To do this highlight EQ:. Then press 2 [×] [α] X [+] 3 [×] {{Y}} [↰] [=] 1 8 [ENTER]. Note that multiplication symbols must be entered when the command line is used while this is not necessary in the Equation Writer.

An entry can be edited on the command line, if necessary. If, for instance, you pressed 5 instead of 8, use [◁] to move the cursor to 5 and press [DEL] 8 to delete 5 and replace it with 8. You can also use the [⇐] to edit your entry. If, for example, you had entered 6 instead of 3 use [⇐] repeatedly to delete 6 and all the entries following it. Then reenter the expression beginning with 3. You can also press [ON] to delete the entire expression and then reenter it. When the editing is complete, press [ENTER] to enter it in the EQ: field. To edit after entering the expression, highlight it by using the arrow keys to scroll to it and select {{EDIT}}.

Once the equation is entered correctly, select a viewing window and then press {{ERASE}} {{DRAW}} to display the graph. You may change the viewing window as desired to reveal more or less of the graph. The window $[-10, 10, -10, 10]$ is shown here.

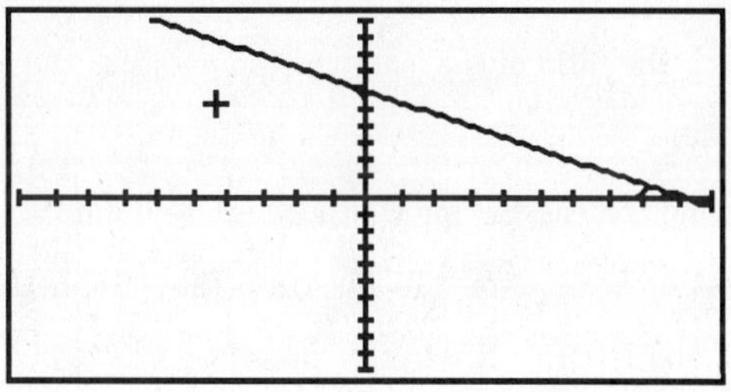

To remove the menu from the display, as seen above, press {{EDIT}} [NXT] {{MENU}}. To restore the menu, press any solid white key.

Example 8(c), page 8: Graph: $x = y^2 + 1$

This is a second-degree equation in the two variables x and y. It is not solved for y, so we will graph it using the Conic type setting. Select TYPE: Conic and INDEP: X as in Example 8(a) above. Then enter the equation. We will use the command line to do this. Press [α] X [↰] [=] {{Y}} [y^x] 2 [+] 1 [ENTER].

Select a viewing window and press {{ERASE}} to clear previous graphs, and the press {{DRAW}} to display the graph. The window shown here is $[-2, 5, -5, 5]$, H-TICK = 1, V-TICK = 1.

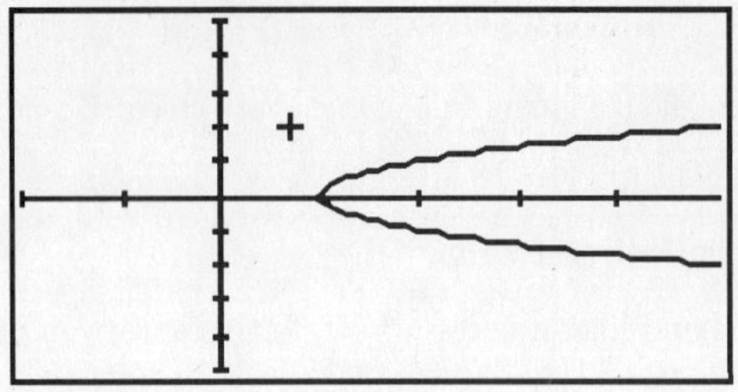

Example 8(c), page 8: Graph: $y = x^4 - 2x^2 - 3$.

This equation is solved for y, so we will graph it using the Function type setting. Open the Plot screen by pressing ↱ PLOT. Highlight TYPE:, press {{CHOOS}}, highlight Function, and press ENTER or {{OK}}. INDEP: should still be set to X. Now enter $x^4 - 2x^2 - 3$ in the EQ: field. To do this, highlight EQ: and press α X y^x 4 $-$ 2 $\times$ {{X}} y^x 2 $-$ 3 ENTER. Then select a window and press {{ERASE}} {{DRAW}}.

SQUARING THE VIEWING WINDOW

In the standard window, the distance between tick marks on the y-axis is 64/131, or about 1/2 the distance between tick marks on the x-axis. It is often desirable to choose window dimensions for which these distances are the same, creating a "square" window. Any window in which the ratio of the length of the y-axis to the length of the x-axis is 1/2 will produce this effect.

This can be accomplished by selecting dimensions for which Ymax $-$ Ymin $= \frac{1}{2}$(Xmax $-$ Xmin). For example, the windows $[-16, 16, -8, 8]$ and $[-8, 8, -4, 4]$ are square. To illustrate this, graph the circle $x^2 + y^2 = 49$ in the window $[-10, 10, -10, 10]$. With TYPE: Conic and INDEP: X selected, press α X y^x 2 $+$ {{Y}} y^x 2 ↰ $=$ 4 9 ENTER. Press {{ERASE}} {{DRAW}} to see the graph.

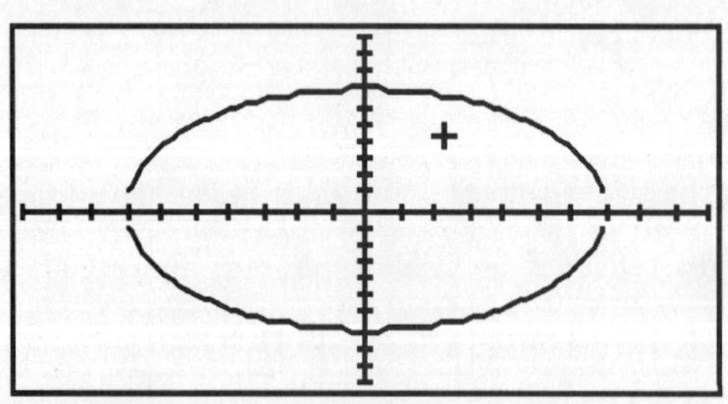

Note that the graph does not appear to be a circle.

Now square the window by changing the window dimensions to $[-16, 16, -8, 8]$, H-TICK = 1, V-TICK = 1, and press PLOT.

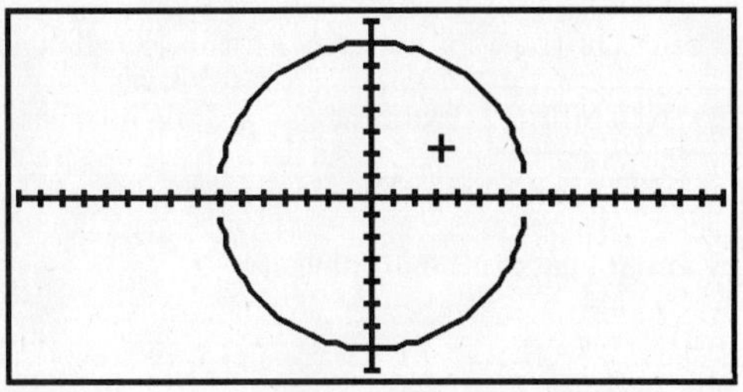

Observe that the distance between tick marks appears to be the same on both axes and that this graph is not distorted.

The window can also be squared by using the default Plot settings, which are $[-6.5, 6.5, -3.1, 3.2]$, H-TICK = 10, V-TICK = 10, PIXELS checked. Do this on the Plot screen as described before by first highlighting any window setting and pressing DEL. Then highlight Reset plot on the message window that appears and press {{OK}} or ENTER. The resulting window dimensions and graph are shown below. Note that the graph appears to be a circle although this window is not a good choice for this particular graph.

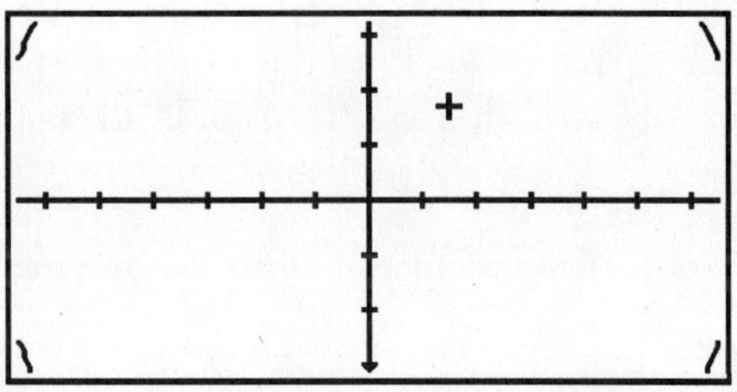

The window can also be squared using the grapher's ZSQR feature. Press ON to return to the Plot setup screen, set the viewing window to any non-square setting such as $[-16, 16, -16, 16]$, and press {{ERASE}} {{DRAW}}. After the graph is drawn, press {{ZOOM}} and then {{ZSQR}}.

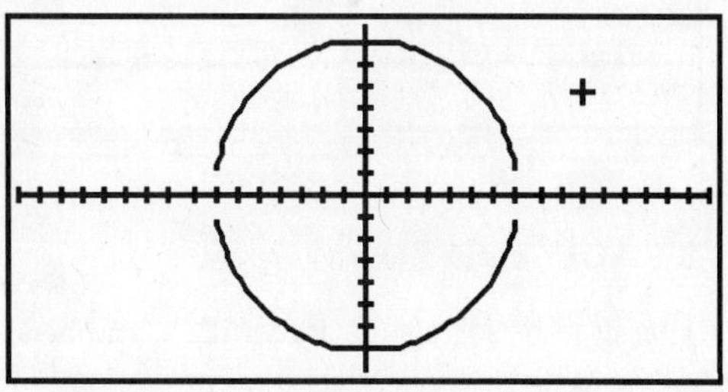

IDENTITIES

An equation that is true for every possible real-number substitution for the variable is an identity. The grapher can be used to provide a partial check whether an equation is an identity.

Example 9 (a), page 11: Determine whether $(x^2)^3 = x^6$ appears to be an identity.

To determine whether this equation appears to be an identity, graph $y = (x^2)^3$ and $y = x^6$ by entering { ' $(x \wedge 2) \wedge 3$ '

'$x\wedge 6$' } in the EQ: field of the Plot screen. On the Plot screen with the EQ: field highlighted, press [↰] [{ }] ['] [↰] [( )] {{X}} [y^x] 2 [▷] [y^x] 3 [▷] ['] {{X}} [y^x] 6 [ENTER]. Examine the graphs in several viewing windows. The graphs appear to coincide no matter what the window. Thus, although there is a possibility that the graphs fail to coincide outside the windows that were examined, the equation appears to be an identity.

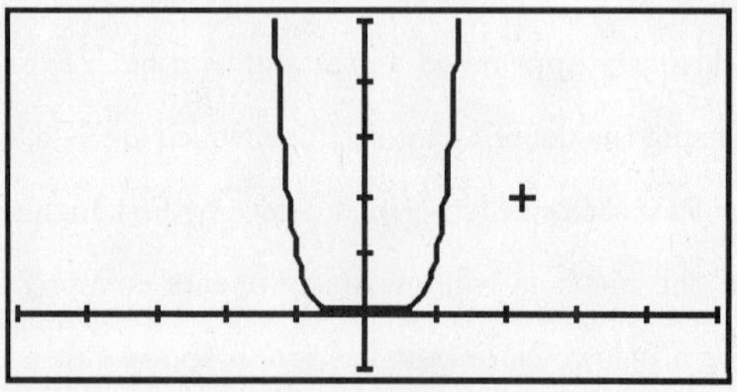

Example 9 (b), page 11: Determine whether $\sqrt{x+4} = \sqrt{x} + 2$ appears to be an identity.

To determine whether this equation appears to be an identity, graph $y = \sqrt{x+4}$ and $y = \sqrt{x} + 2$ as { '$\sqrt{}(x+4)$' '$\sqrt{}x + 2$' } in the EQ: field of the Plot screen. Press [↰] [{ }] ['] [$\sqrt{x}$] [↰] [( )] {{X}} [+] 4 [▷] [▷] ['] [$\sqrt{x}$] {{X}} [+] 2 [ENTER]. Note that parentheses must be used when entering $\sqrt{x+4}$. Without them the expression entered would be $\sqrt{x} + 4$.

Any window that includes a portion of the first quadrant will show that the graphs differ. Thus, the equation is not an identity.

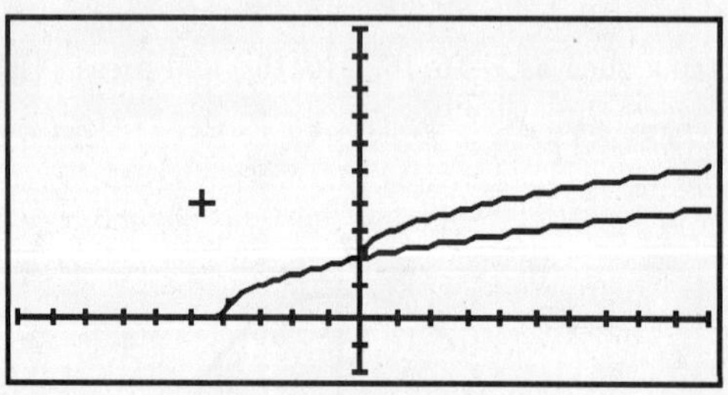

SOLVING EQUATIONS USING TRACE AND ZOOM

There are several techniques that can be used to solve equations with a grapher. One such technique uses the grapher's Trace and Zoom features.

Example 10, page 13: Solve $x^3 - 3x + 1 = 0$. Approximate the solutions to three decimal places.

The solutions of this equation are the first coordinates of the x-intercepts of the graph of $y = x^3 - 3x + 1$. To find these coordinates we first graph $y = x^3 - 3x + 1$ in a viewing window that shows all of the x-intercepts. Enter the equation on the Plot setup screen with TYPE: Function and INDEP: X selected. Highlight EQ: and press [α] X [y^x] 3 [−] 3 [×] {{X}} [+] 1 [ENTER] and select the window $[-10, 10, -10, 10]$. Note: The Equation Writer can also be used to enter the function.

To do this go to Plot setup, highlight EQ:, and press [⬑] [EQUATION] [α] X [y^x] 3 [▷] [−] 3 [α] X [+] 1 [ENTER]. Press {{ERASE}} {{DRAW}} to see the graph.

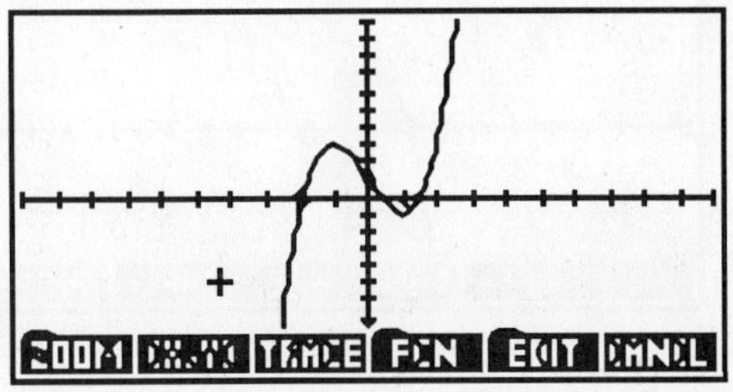

We see that x-intercepts occur near $x = -2$, $x = 0$, and $x = 2$. A portion of the viewing window can be enlarged near each of these values in order to find the desired three decimal place approximation. For example, let's examine the graph near $x = 0$.

Press {{TRACE}} to activate trace mode. Note that a small box appears in the trace menu selection when trace mode is activated.

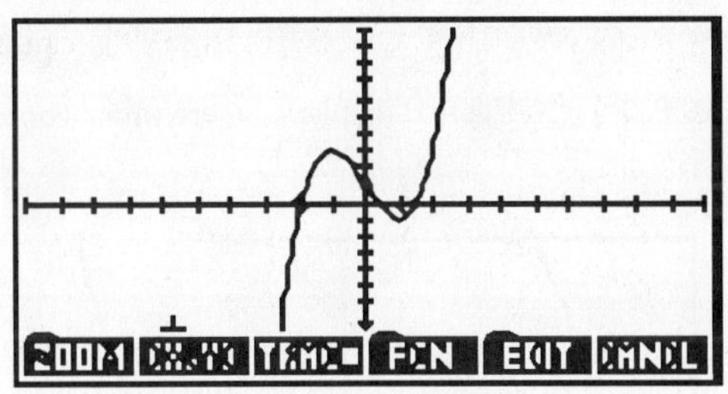

Then press {{(X,Y)}} to see the cursor coordinates on-screen. The X- and Y-values at the bottom of the screen indicate the coordinates of the point where the cursor is positioned. Note that non-integer coordinates are expressed in scientific notation. For instance, the coordinates X: 3.08E−1 and Y: 1.06E−1 represent the point $(3.08 \times 10^{-1}, 1.06 \times 10^{-1})$, or (0.308,0.103). (See Section R.2 of the text for a discussion of scientific notation.) To move the Trace cursor along the graph, use [◁] and [▷], and if there is more than one graph, use [△] and [▽] to move from graph to graph. Note that the TRACE cursor always remains on the curve. Thus, even when the cursor appears to be off the graph, when TRACE mode is activated the coordinates on the screen are actually on the graph. In the figure below, for example, we see that when $x = -4$, the function value is -51.

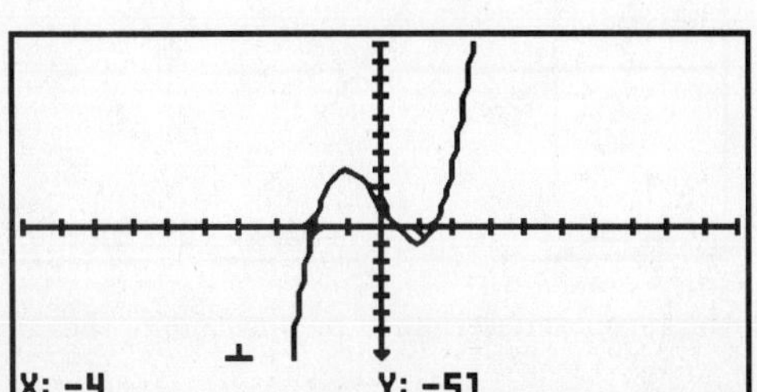

To see the equation of the curve being traced, press any solid white key, and then press {{FCN}} NXT {{VIEW}}; the equation will be displayed briefly.

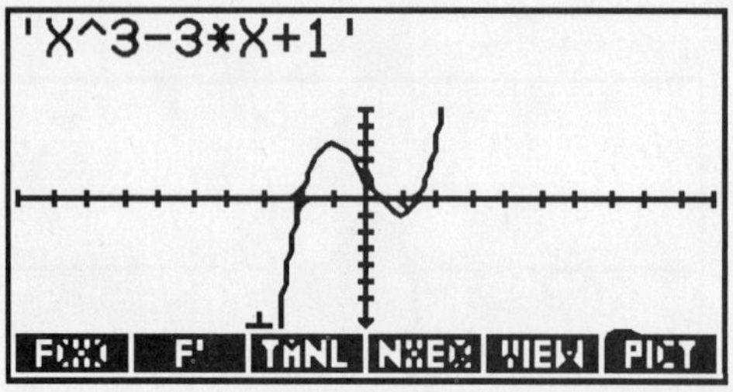

To return to the previous display, press {{PICT}} {{TRACE}} {{(X,Y)}}.

In order to find the middle x-intercept, we enlarge or zoom in on the portion of the graph near $x = 0$ by first positioning the cursor as close to this intercept as possible. That is, near $x = 0$, position the cursor as close as possible to $y = 0$. Then press any solid white key, {{ZOOM}}, and {{ZIN}} to zoom in on the curve at the point where the cursor is positioned. If the zoom factors are not set to 4×4, which we use in this example, press {{ZOOM}} and then press {{ZFACT}}. Set H-FACTOR and V-FACTOR to 4 and then press {{OK}} {{PICT}}. If you wish to have the graph centered on the cursor crosshairs when zooming, check RECENTER AT CROSSHAIRS. This can also be done from the menu on-screen by pressing {{CNTR}}. Press {{TRACE}} {{(X,Y)}} to display the crosshair coordinates again.

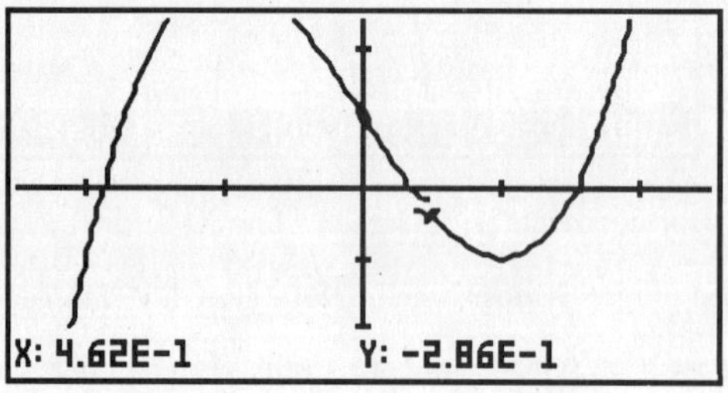

To enlarge this portion of the graph further move the cursor as close as possible to $y = 0$ near $x = 0$. Now press any solid white key, {{ZOOM}} {{ZIN}}, and then press {{TRACE}} {{(X,Y)}} .

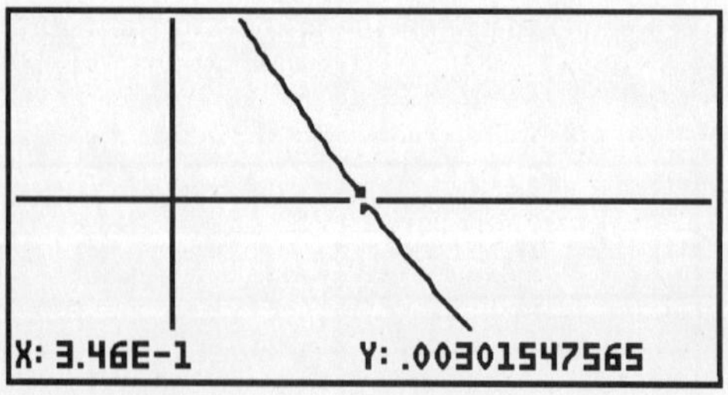

Continue tracing and zooming in until the x-values determined by positioning the cursor just to the left and just to the right of the x-intercept are the same when rounded to three decimal places. We find that $x \approx 0.347$ at the middle x-intercept.

This is one solution of the equation $x^3 - 3x + 1 = 0$.

To find a second solution of the equation return to the Plot screen and enter the original window settings, or press {{ZOOM}} {{ZDFLT}} to use the default window; then trace to a position on the curve near another x-intercept and zoom in as described above. Continue to trace and zoom in until the desired accuracy is obtained. Repeat this process to find the third solution. The other two solutions are about -1.879 and 1.532.

At any point in this process, we can zoom out to the previous window using Zoom-to-Last by pressing {{ZOOM}} {{NXT}} {{NXT}} {{ZLST}}. Only the previous window is retained in memory, however, so we can only zoom out one time using Zoom-to-Last.

The Zoom-to-Box operation can also be used to enlarge a portion of the viewing window. We select diagonally opposed corners of a box that defines a new window. For example, graph $y = x^3 - 3x + 1$ in the window $[-10, 10, -10, 10]$. Then use the arrow keys to move the cursor to any corner of the box to be defined. Note that the cursor used in this operation is a free-moving cursor. That is, it can be positioned at any point on the screen as opposed to the trace cursor whose position is restricted to points on the graph.

Press {{ZOOM}} {{BOXZ}} to select the first corner. Now use [◁], [▷], [△], and/or [▽] to move the cursor to the corner of the box diagonally opposite this one. As you move the cursor away from the first corner the boundaries of the box appear on the screen and change as the position of the cursor changes.

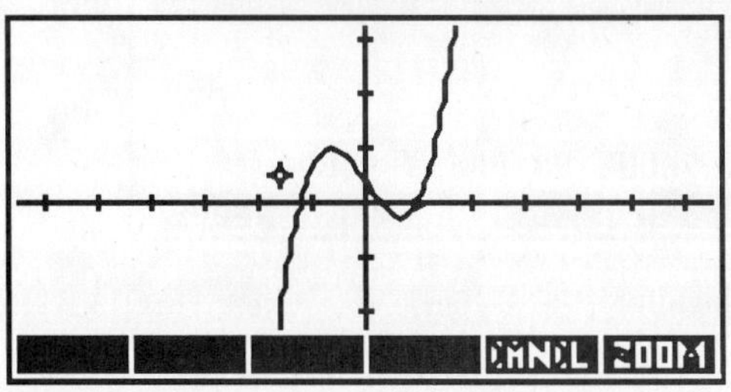

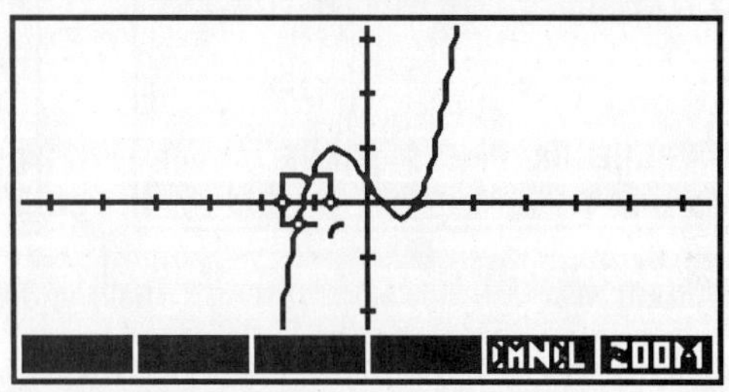

When the box is defined as you want it, press {{ZOOM}} to show only the portion of the graph in the box.

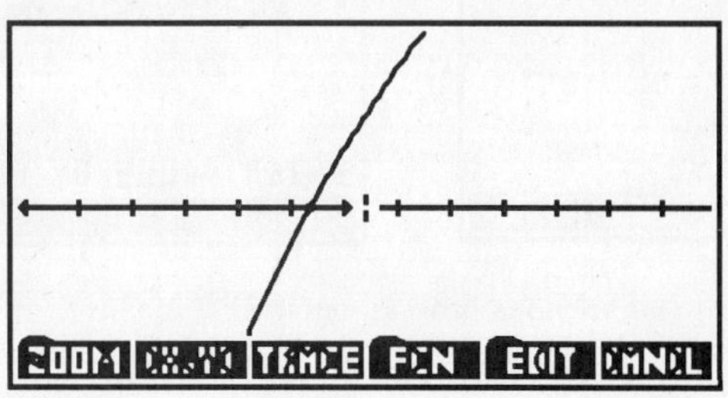

Now we can zoom in or use Zoom-to-Box again to further enlarge a portion of the graph, if desired.

SOLVING EQUATIONS USING SOLVE OR SOLVER

The Solve application can also be used to solve equations.

Example 10, page 13: Solve $x^3 - 3x + 1 = 0$. Approximate the solutions to three decimal places.

Begin as in the procedure described above for using Trace and Zoom by graphing $y = x^3 - 3x + 1$ in a window that shows all of the x-intercepts. This shows us how many real-number solutions the equation has and their approximate values. We see as before that the x-intercepts occur near $x = -2$, $x = 0$, and $x = 2$. We will first find the solution near $x = -2$.

Press {{CANCL}} or [ON] to go back to the Plot screen. Then access the Solve application and choose Solve equation ... by pressing [↱] [SOLVE] {{OK}} (or [ENTER]). The EQ: field is highlighted and the equation entered in the Plot setup screen appears there.

We can now enter an estimated value, or initial guess, for the solution in the X: field. An initial guess is particularly important in the event that there is more than one solution. Since we are seeking the solution near $x = -2$, we use this for the guess. Position the cursor beside X: and enter -2 as the initial guess using the keystrokes 2 [+/−] [ENTER]. Press [▽] to highlight the X: field and then press {{SOLVE}}. The grapher returns the value -1.87938524157. Thus one solution is about -1.879.

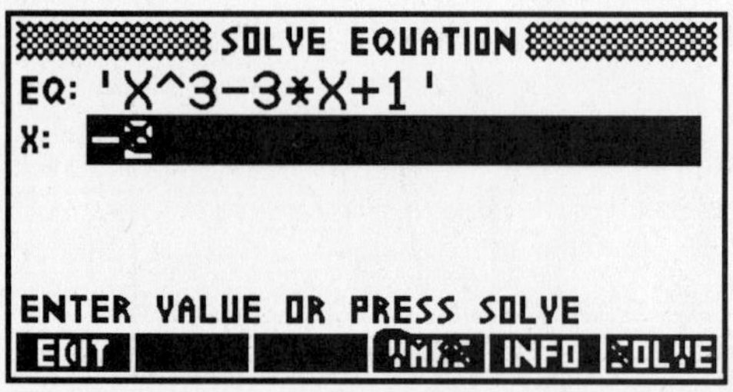

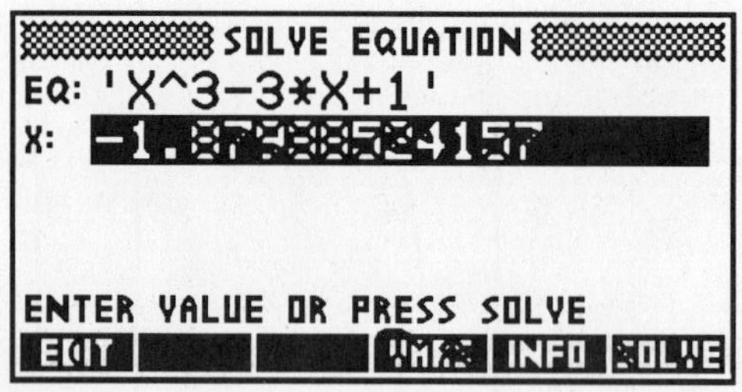

To find the solution near $x = 0$ enter 0 in the X: field and then highlight the X: field. Now press {{SOLVE}} and the value .347296355335 is returned.

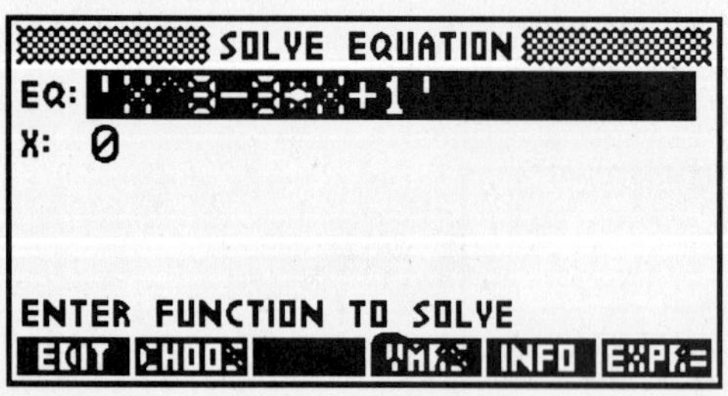

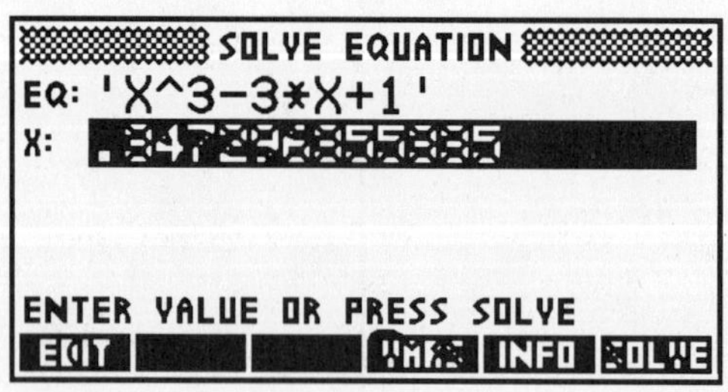

To find the solution near $x = 2$, repeat the previous process, entering 2 in the X: field.

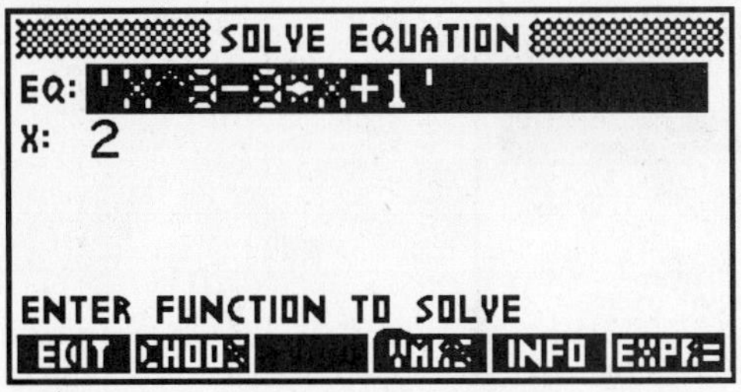

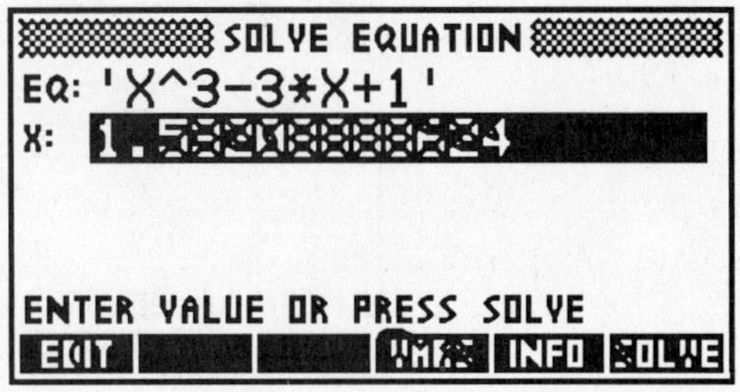

Thus, as we saw before, the other two solutions are about 0.347 and 1.532.

After finding a solution with Solve, more information about the nature of the solution can be obtained by pressing {{INFO}}. One of three messages will appear:

1) Zero. This indicates that Solve found a point, to an accuracy of 12 digits, where the equation's value is zero.

2) Sign Reversal. This indicates that Solve found two points where the value of the equation has opposite signs. However, it could not find a point between those two points where the value of the equation is zero. Either the two points are too close together (they differ by one in the twelfth digit) or the equation is not real-valued between the points. Solve returns the point where the equation's value is closer to zero.

This can happen at a legitimate solution, but it can also occur at a value of the variable for which the function is not defined. To decide which is the case, we can evaluate the function at the Sign Reversal point. To do this remove the information message by pressing {{OK}}, highlight the EQ: field on the Solve Equation screen and then press {{EXPR =}}. Now press [ON] to return to the stack on the Home screen. If the value beside 1: Expr. is not close to 0, then the value found using Solve is not a solution of the equation. This procedure can be avoided however, if the function has been graphed, telling us how many solutions the equation has and their approximate values.

3) Extremum. This indicates that Solve either found a point where the equation has a local maximum or minimum or it stopped searching at the largest or smallest number the grapher can represent, ±9.99999999999E499.

The message "Bad Guess(es)" means no solution was found because one or more of the initial guesses are not in the equation's domain, such as an initial guess of -1 for $\sqrt{x}$. The message "Constant?" means no solution was found because the equation has the same value at every point sampled.

The process of searching for a zero can be observed by pressing any key except [ON] (which will cancel the operation) immediately after pressing {{SOLVE}}. Two intermediate guesses and, on the left, the signs of the equation's value when evaluated at those guesses, will be displayed. If no convergence in process is observed, cancel the operation by pressing [ON] and start over but use a different initial guess.

Example 12, page 15: Solve: $\frac{2}{3}x - 7 = 5$.

Press [↱] [SOLVE] [ENTER] (or {{OK}}). The Solve Equation view will appear with the EQ: field highlighted. Enter the equation $\frac{2}{3}x - 7 = 5$ by pressing 2 [÷] 3 [×] [α] X [−] 7 [↰] [=] 5 [ENTER].

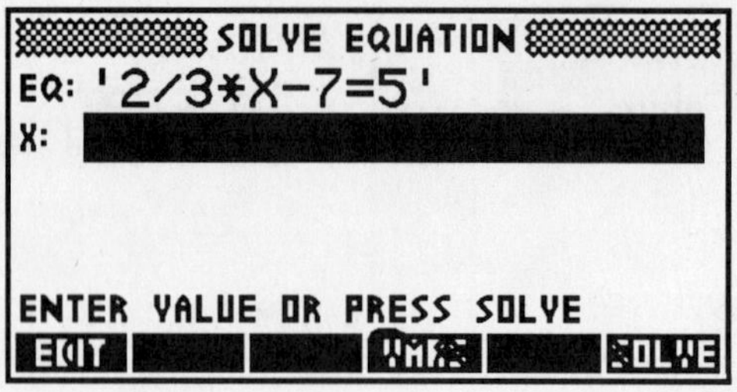

Use the Plot view to find an initial guess for x. First go to Plot by pressing [↱] [PLOT] and set the viewing window. Here we use $[-5, 35, -10, 10]$, H-TICK = 5, V-TICK = 2, PIXELS unchecked. Now press {{OK}} {{ERASE}} {{DRAW}}, and note that $y = \frac{2}{3}x - 7$ and $y = 5$ are graphed. The x-coordinate of their point of intersection is the solution of the original equation. Move the cursor as close to the intersection as possible and press {{X,Y}} to see the coordinate of the point where the cursor is positioned. The x-value at that point will be an initial guess for x. We could also find an initial guess by simply observing from the graph that the x-coordinate of the point of intersection is approximately 20. This could also serve as the initial guess for x.

Press {{CANCL}} [↱] [SOLVE] [ENTER] to return to the Solve Equation view. Then press [▽] to highlight the X: field and enter the guess found from the graph followed by [ENTER].

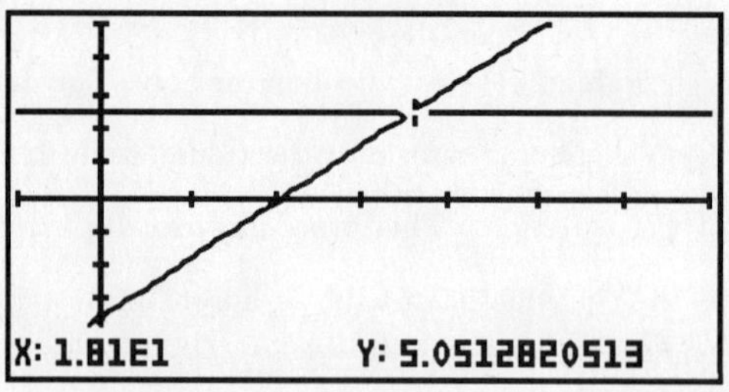

Now highlight the X: field again and press {{SOLVE}}. The grapher returns the solution of the equation, 18

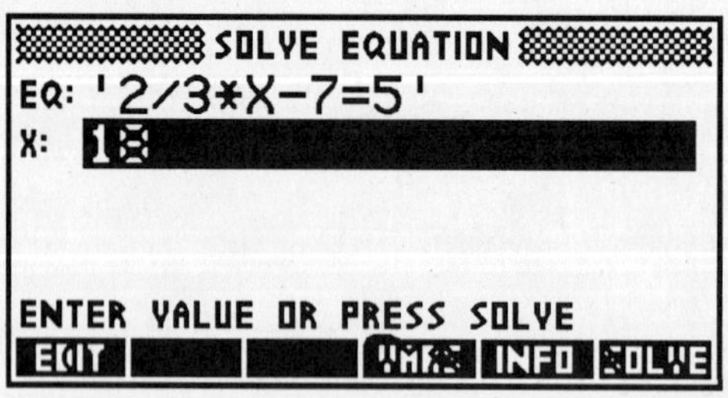

SOLVING EQUATIONS USING ROOT

A solution of an equation with 0 on one side is usually called a zero, but is sometimes called a root. We can find the zeros of an equation on the HP 48G by using the ROOT operation.

Example 10, page 13: Solve $x^3 - 3x + 1 = 0$. Approximate the solutions to three decimal places.

Graph $y = x^3 - 3x + 1$ in the standard window as before. To find the solution near $x = 0$, first move the cursor as close to the x-intercept near $x = 0$ as possible. Press {{FCN}}. Choose {{ROOT}}. The x-value that the grapher returns is the solution. We see again that the solution is about 0.347.

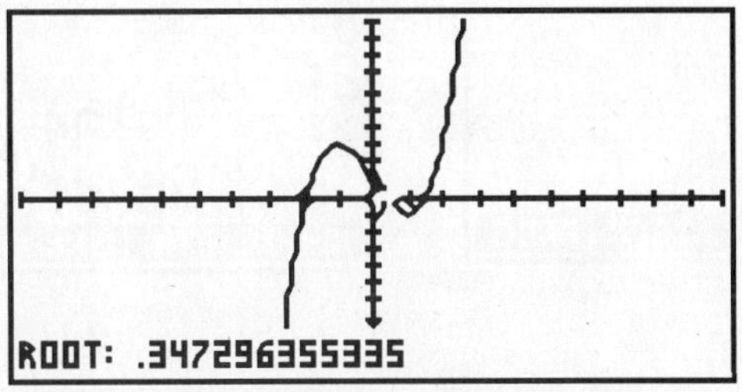

Press any white key to return to the screen with menus at the bottom. Then to find the solution near $x = 2$, move the cursor as close as possible to the x-intercept near $x = 2$ and press {{ROOT}}. Repeat this procedure to find the solution near $x = -2$.

FINDING POINTS OF INTERSECTION

There are several ways in which the grapher can be used to determine the point(s) of intersection of two graphs.

Example 11, page 14: Find the point of intersection of the graphs of the equations $y_1 = 3x^5 - 20x^3$ and $y_2 = 34.7 - 1.28x^2$. Approximate the coordinates to three decimal places.

First we will enter the functions in the Plot view. They can be entered directly in a list in the Plot view by pressing [⮢] [{ }] ['] 3 [×] {{X}} [y^x] 5 [−] 2 0 [×] {{X}} [y^x] 3 [◁] ['] 3 4 [.] 7 [−] 1 [.] 2 8 [×] {{X}} [y^x] 2 [ENTER]. The Variable Browser can also be used as follows. First, choose a directory in which you will store the functions as variables. For example, for the Home directory press [⮣] [HOME]. (HOME is the right-shift operation associated with the ['] key.)

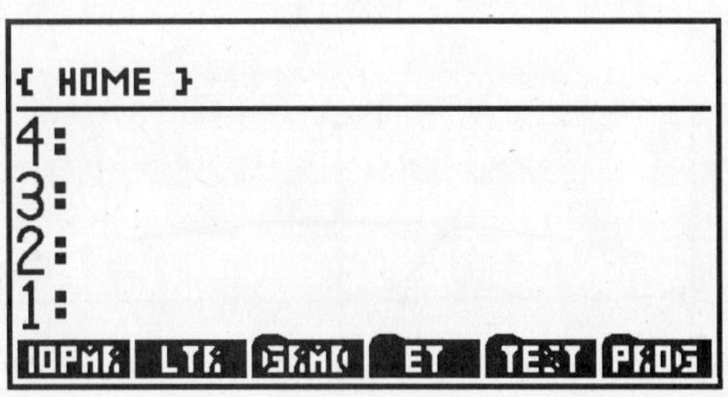

Your Home directory may look different from the one in the figure above due to different variables you may have created. Either the Equation Writer or the command line of the Home screen can be used to enter y_1 and y_2 in the Home directory. We will illustrate by using the Equation Writer to enter y_2 and the command line to enter y_1.

To open the Equation Writer press [⮢] [EQUATION]. Then press 3 4 [.] 7 [−] 1 [.] 2 8 [α] X [y^x] 2 to enter y_2 and finally press [ENTER] to place y_2 on the stack. Now enter y_1 on the command line of the Home screen. Note that when the command line is used, the ' ' delimiters and multiplication symbols must be used. This is not necessary in the Equation

Writer. Press ' 3 × α X y^x 5 − 2 0 × α X y^x 3. Again, press ENTER to place y_1 on the stack.

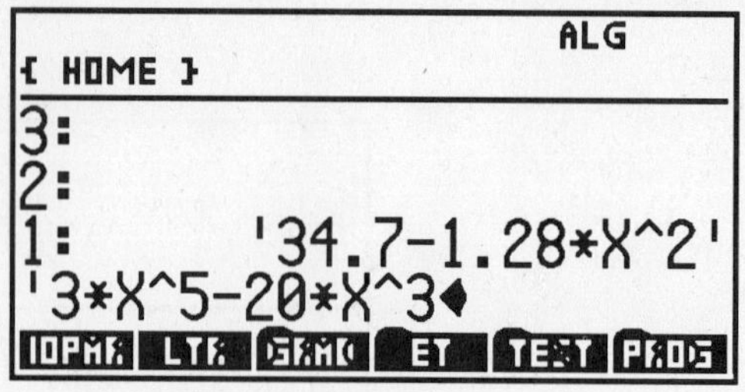

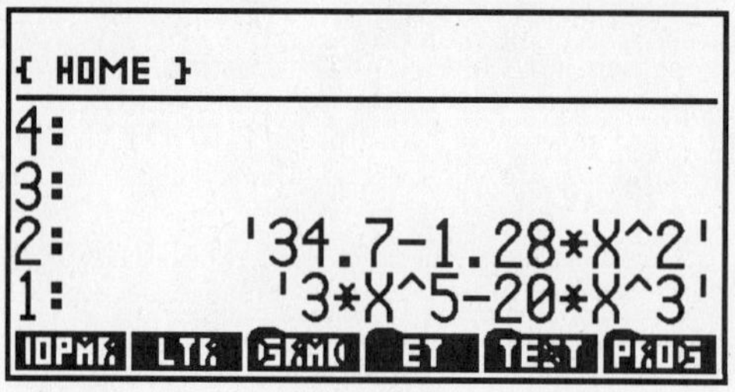

Now press ' α F 1 STO ' α F 2 STO to store y_1 as F1 and y_2 as F2. Note that the functions now no longer appear on the Home screen, but they are now stored as the variables F1 and F2 in the Home directory. F1 and F2 will now also appear as menu items on the Home screen. If they are not currently visible it might be necessary to press VAR and/or NXT in order to see them. In fact, it might be necessary to press NXT more than once.

Now open the Plot screen, highlight the EQ: field, and press {{CHOOS}}. F1 and F2 are in the message window that appears; use ▽ to scroll to each of them in turn and checkmark each by pressing {{√ CHK}}. Then press ENTER. Select a window that shows all of the points of intersection. We use $[-4, 4, -80, 80]$, H-TICK = 1, V-TICK = 20, PIXELS unchecked, here. Now press {{ERASE}} {{DRAW}} to graph F1 and F2.

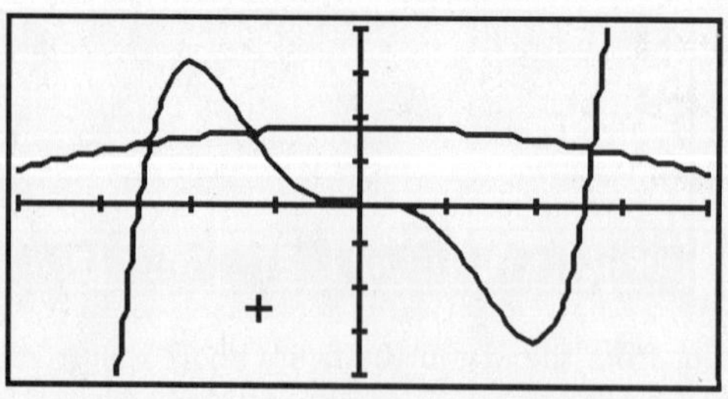

There are three points of intersection. One way to find their coordinates is to use Trace and Zoom. (See "Solving Equations Using Trace and Zoom" on page 10 of this manual.) Another method is to use the Intersection operation. We will illustrate its use by finding the coordinates of the point of intersection at the far left. Move the cursor to that point. Then press {{FCN}} {{ISECT}}. The grapher returns the coordinates of the point of intersection. We see that they are about $(-2.463, 26.936)$.

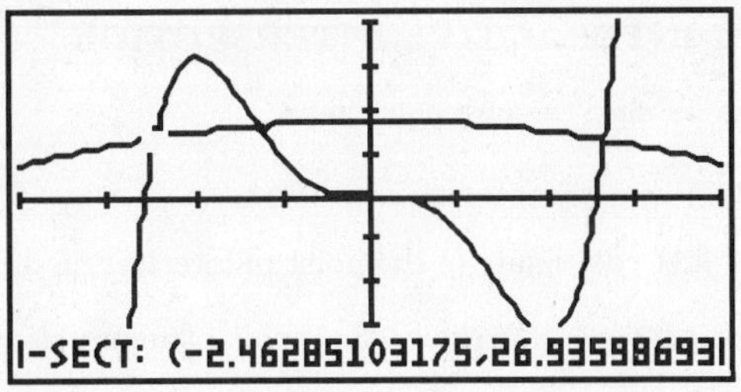

The coordinates of the other two points of intersection can be found by using the Intersection operation two more times. Press any white key each time to restore the screen menu.

The Solve application can also be used to find the x-coordinates of the points of intersection by finding solutions of F1 = F2, or F1 − F2 = 0. Again it is necessary to graph F1 and F2 as we have just done in order to find guesses for the solutions. From the graph we see that we can use −2.5 as an initial guess for the left-most solution. Enter the Solve equation . . . application by pressing [↱] [SOLVE] [ENTER]. Enter F1 = F2 in the EQ: field by pressing [α] F 1 [↰] [=] [α] F 2 [ENTER]. Then enter −2.5 in the X: field. Highlight the X: field and press {{SOLVE}}. The grapher returns the value −2.46285103175. This is the first coordinate of the point of intersection.

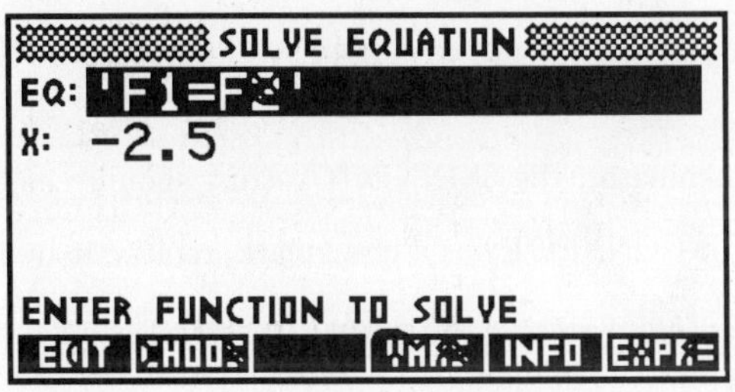

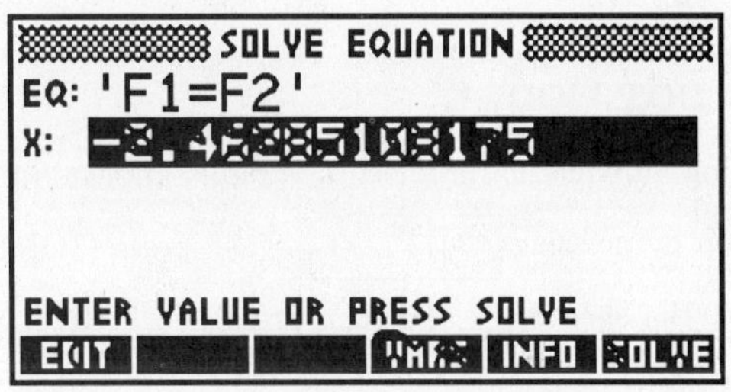

To find the second coordinate, highlight the EQ: field. Enter F1 = Y or F2 = Y. For example, enter F1 = Y by pressing [α] F 1 [↰] [=] [α] Y [ENTER]. Then highlight Y: and press {{SOLVE}}. The value 26.935986939 is returned.

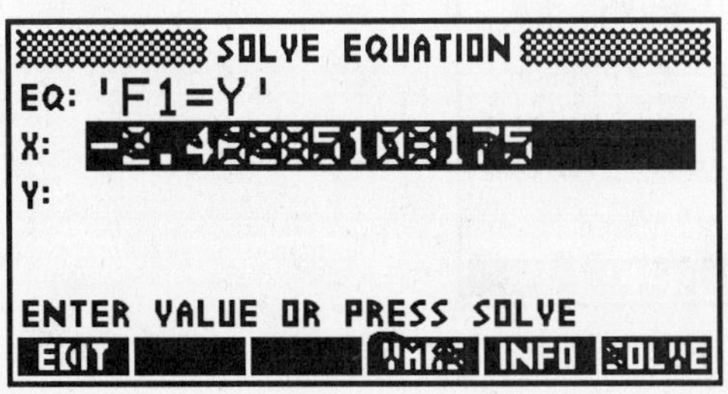

Thus, we see that the coordinates of the point of intersection are about $(-2.463, 26.936)$. If we had used the rounded X-value of −2.463 in F1 to compute the corresponding Y-value, the result would have been 26.907983054. If we had used the rounded X-value in F2, the resulting Y-value would have been 26.93504768. To achieve the greatest accuracy, then, it is important to use the unrounded X-value in this computation.

SOLVING EQUATIONS USING INTERACTIVE INTERSECTION

The Intersection operation can also be used to solve equations.

Example 12, page 15: Solve: $\frac{2}{3}x - 7 = 5$.

The solution of this equation is the first coordinate of the point of intersection of the graphs of $F1 = \frac{2}{3}x - 7$ and $F2 = 5$. Thus, we graph these equations in a window that shows the point of intersection and use Intersection as described in "Finding Points of Intersection" on page 17 of this manual. The x-coordinate of the point of intersection is 18. This is the solution of the equation.

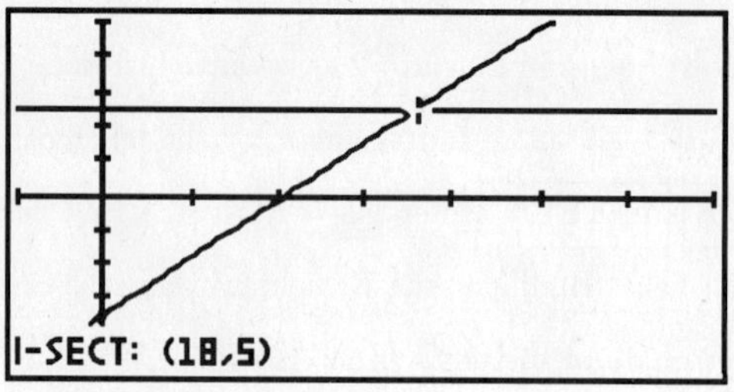

SELECTING DOT MODE

When graphing an equation in which a variable appears in a denominator, the CONNECT setting should not be used. Enter Plot Setup by pressing [↱] [PLOT]. Then press {{OPTS}}. If CONNECT has a checkmark, remove it by pressing [▽] to highlight the checkmark and then press {{√ CHK}}. This graphing mode is referred to as DOT mode on some graphers.

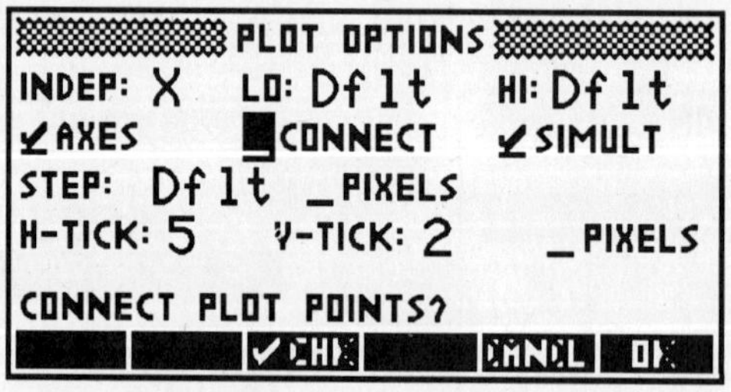

See page 16 in the text for further explanation of why DOT mode is used in certain cases and for an illustration of an equation graphed in both CONNECT and DOT modes.

Chapter 1
Graphs, Functions, and Models

FINDING FUNCTION VALUES

When a formula for a function is given, function values for real-numbered inputs can be found in several ways on a grapher.

Example 4 (a), (b), page 23: For $f(x) = 2x^2 - x + 3$, find $f(0)$ and $f(-7)$.

Method 1: Substitute the inputs directly in the formula. For example, to find $f(0)$ on the Home screen using reverse Polish notation press 2 [ENTER] 0 [↰] [x^2] [×] 0 [−] 3 [+]. To find $f(-7)$ using algebraic notation press ['] 2 [×] [↰] [x^2] [+/− 7] [▷] [−] [+/−] 7 [+] 3 [EVAL].

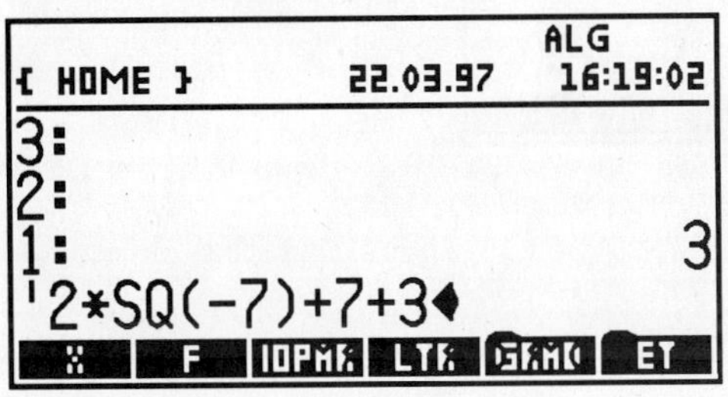

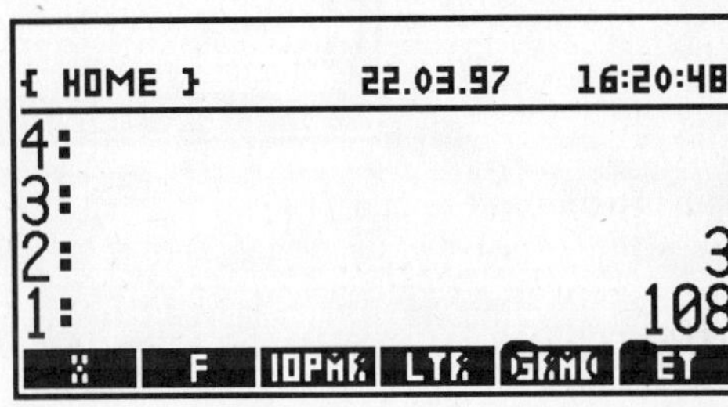

To use the Equation Writer and algebraic notation press [↰] [EQUATION] 2 [↰] [( )] [+/−] 7 [▷] [y^x] 2 [▷] [−] [+/−] 7 [+] 3 [ENTER] [EVAL].

Method 2: Using the Equation Writer or the command line of the Home screen, enter $2x^2 - x + 3$ on the stack. Remember that the delimiters ' ' must be used if the expression is entered on the command line. Pick a variable name, such as F, in which to store this expression. and press ['] [α] F [STO]. {{F}} should now appear as a menu selection. It might be necessary to press [VAR] and/or [NXT] in order to see it. In fact, it might be necessary to press [NXT] more than once. Now store 0 as the variable X. Press 0 [ENTER] ['] [α] X [STO]. Now press {{F}} [EVAL] to evaluate F at $x = 0$. To evaluate F at $x = -7$, store -7 as X by pressing 7 [+/−] [ENTER] ['] {{X}} [STO], and the repeat the process above for evaluation, pressing {{F}} [EVAL].

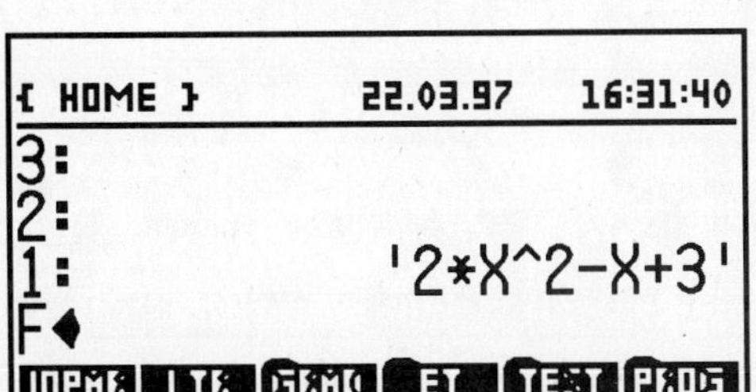

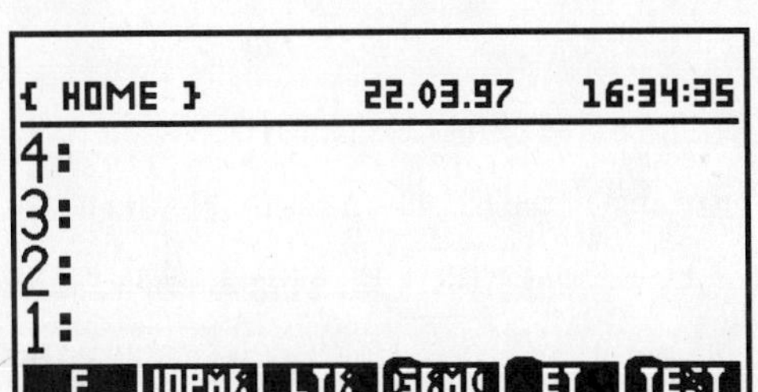

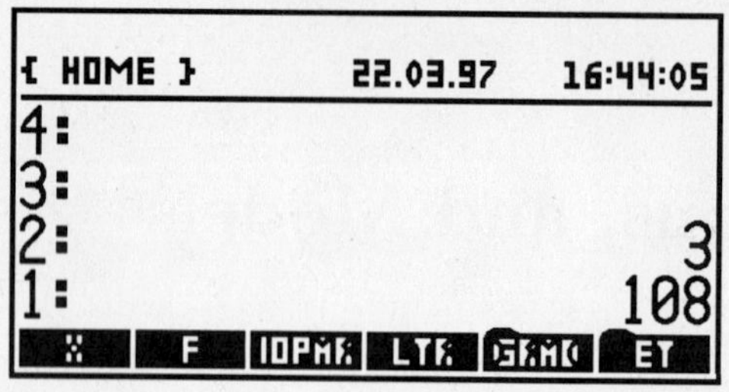

Method 3: The Solve application can be used to find function values. Press ↱ SOLVE ENTER. Enter $2x^2 - x + 3$ in the EQ: field. Then with the X: field highlighted, press 0 ENTER {{EXPR =}} to find $f(0)$. Nothing will seem to have happened, but press ON to return to the stack; there you will see the result.

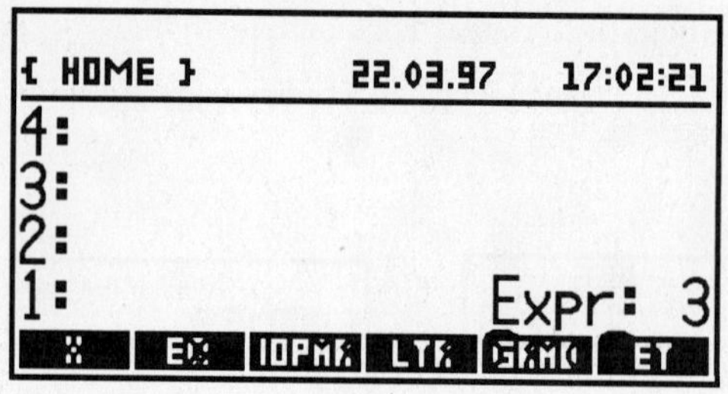

A similar procedure is performed to find $f(-7)$.

GRAPHING FUNCTIONS DEFINED PIECEWISE

For some plot types, the HP 48G can plot programs. A program can be plotted if it uses nothing from the stack, contains the independent variable, and returns exactly one real number (or one complex number for parametric plots). We use a program for graphing piecewise functions.

Example 4, page 38: Graph

$$f(x) = \begin{cases} 4, & \text{for } x \leq 0, \\ 4 - x^2, & \text{for } 0 < x \leq 2, \\ 2x - 6, & \text{for } x > 2. \end{cases}$$

Enter the Plot setup, choose Function as the plot type, and enter the following program in the EQ: field:

```
<< CASE
    X 0 ≤ THEN 4 END
    X 2 ≤ THEN '4 – X∧2' EVAL END
    '2 * X – 6' EVAL
END >>
```

Note that all programs in the HP 48G must use the angle-bracket delimiters << >>. As long as commands are separated by spaces, created by pressing SPC, the program may be entered on a single line. As before press ↱ CHARS to see the ≤ symbol, highlight it, and press {{ECHO}} ENTER to include it in the program. The alpha function can be locked in by pressing α α. This will allow you to type consecutive alpha characters without pressing α before each one. Press

$\boxed{\alpha}$ again to unlock this function. Plot as usual after doing this.

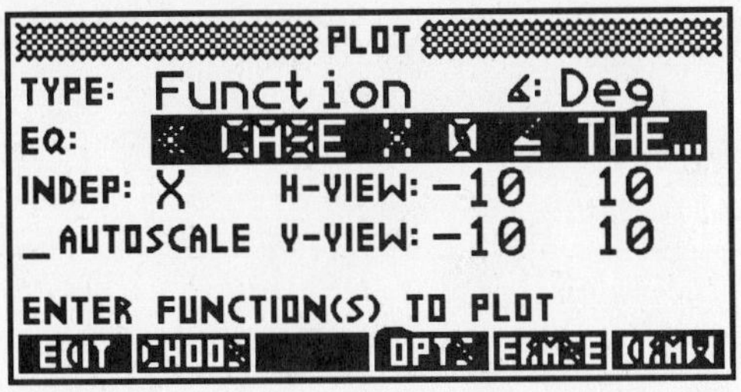

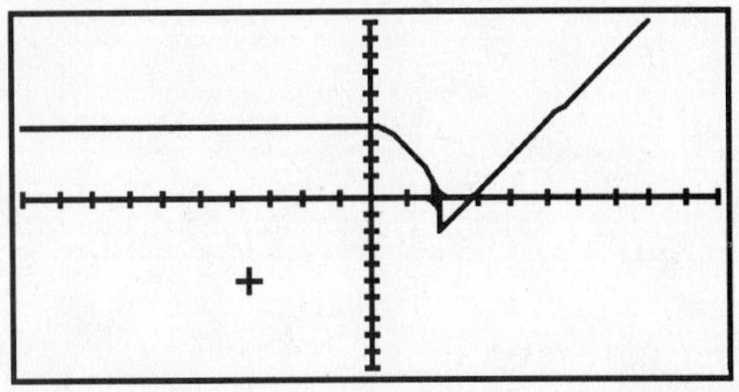

For extensive details about programming the HP 48G, consult the *HP 48G Series Advanced User's Reference Manual*, HP Part No. 00048-90136.)

The greatest integer function can be graphed with the following program: << X FLOOR >>. Be sure that CONNECT is unchecked in the Plot Options.

THE ALGEBRA OF FUNCTIONS

The grapher can be used to evaluate and graph combinations of functions.

Example 1 (b), page 92: Given that $f(x) = x + 1$ and $g(x) = \sqrt{x+3}$, find $(f+g)(6)$.

Store $f(x)$ and $g(x)$ into the variables F and G, respectively, on the stack. Open the Solve application, choose Solve equation . . . , and enter F + G in EQ:. Enter 6 in X: and press {{EXPR =}} $\boxed{\text{ON}}$ $\boxed{\text{EVAL}}$. We find that $(f+g)(6) = 10$.

To view the graphs of F, G, and F + G, enter the list {F G 'F + G'} in the EQ: field of the Plot view. TYPE: should be set to Function. To see only F + G, press {{EDIT}} and use $\boxed{\text{DEL}}$ to delete everything but 'F + G'.

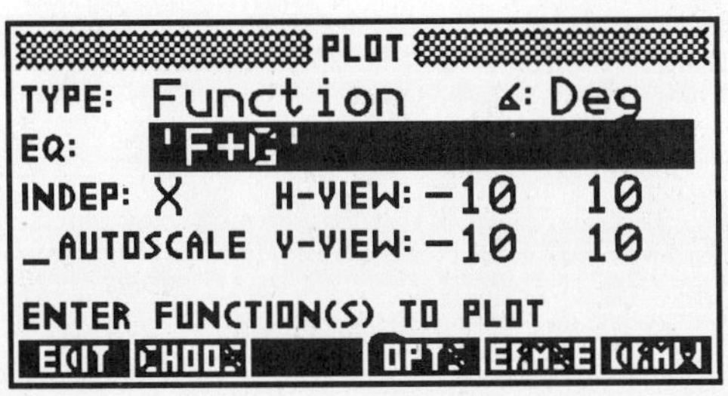

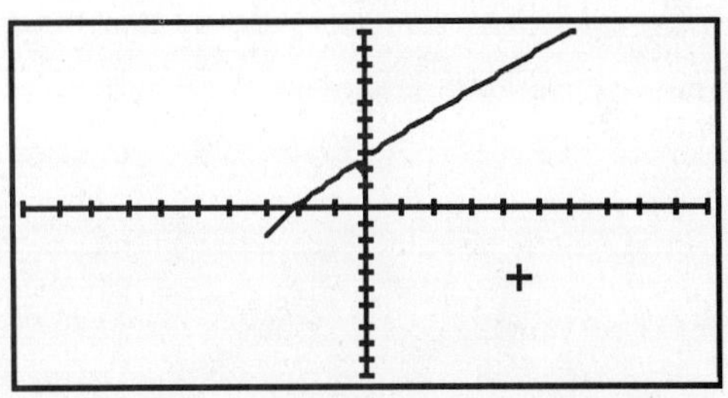

Chapter 2
The Trigonometric Functions

CONVERTING BETWEEN D°M'S″ AND DECIMAL DEGREE MEASURE

The Real-Number functions HMS→ and →HMS can be used to convert D°M'S″ notation to decimal notation and vice versa.

Example 5, page 122: Convert 5°42′30″ to decimal degree notation.

Press [↰] 4 [NXT]. Now enter 5°42′30″ as the number of degrees followed by a decimal point and then the remainder of the digits: 5.4230. Press {{HMS→}}. The grapher returns 5.708333333, so 5°42′30″ ≈ 5.71°.

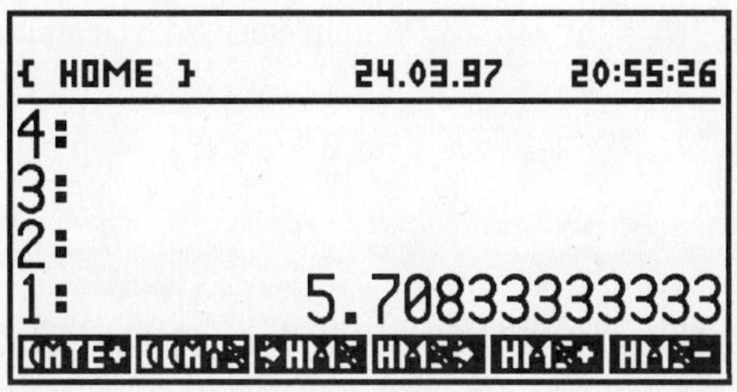

Example 6, page 122: Convert 72.18°to D°M'S″ notation.

Press [↰] 4 [NXT]. Now press 7 2 [·] 1 8 {{→HMS}}. The grapher returns 72.1048, where the digits in front of the decimal point are degrees, the first two digits after the decimal point are minutes, and the two final digits are seconds. Thus we have 72°10′48″.

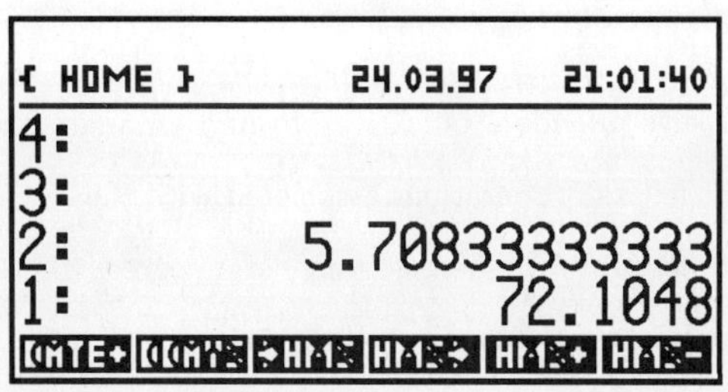

The procedure for converting D°M'S″ notation to decimal notation requires that both the minutes and the seconds be expressed using two digits. For example, enter 8°5'3" as 8.0503.

FINDING TRIGONOMETRIC FUNCTION VALUES

The grapher's SIN, COS, and TAN operations can be used to find the values of trigonometric functions. When angles are given in degree measure, the grapher must be set in Degree mode. To do this press [↱] [MODES], highlight the Angle Measure: field, press {{CHOOS}}, highlight Degrees, and press {{OK}} {{OK}}.

Example 7, page 122: Find the trigonometric function value, rounded to four decimal places, of each of the following.

a) tan 29.7° b) sec 48° c) sin 84°10′39″

a) With the grapher set in Degree mode, press 2 9 [·] 7 [TAN]. We find that tan 29.7° $\approx$ 0.5704.

b) The secant, cosecant, and cotangent functions can be found using the reciprocals of the cosine, sine, and tangent functions, respectively. This can be done either by entering the reciprocal or by using the [$1/x$] key. To find sec 48° we can enter the reciprocal of cos 48° by pressing 1 [ENTER] 4 8 [COS] [÷]. To find sec 48° using the [$1/x$] key press 4 8 [COS] [$1/x$]. The result is sec 48° $\approx$ 1.4945.

The secant, cosecant, and cotangent functions can also be implemented as simple programs:

SEC: <<COS INV>>

CSC: <<SIN INV>>

COT: <<TAN INV>>

For example, to save the secant program in the variable SEC, press [↰] [<< >>] [COS] [$1/x$] [ENTER] ['] [α] [α] S E C [α] [STO]. The keystrokes [α] [α] I N V could be used instead of [$1/x$]. SEC should now appear as a menu selection. If it does not, press [VAR] and/or [NXT]. In fact, it might be necessary to press [NXT] more than once before SEC appears on the menu. To run the program, enter a number and press {{SEC}}.

c) First convert 84°10′39″ to decimal form, and then press [SIN]. We find that sin 84°10′39″ $\approx$ 0.9948.

FINDING ANGLES

The inverse trigonometric function keys provide a quick way to find an angle given a trigonometric function value for that angle. These keys are actually the left-shift operations associated with the corresponding trigonometric function keys.

Example 8, page 123: Find the acute angle, to the nearest tenth of a degree, whose sine value is approximately 0.20113.

We use the inverse sine key to approximate this angle. With the grapher set in Degree mode press [·] 2 0 1 1 3 [↰] [ASIN]. We find that the desired acute angle is approximately 11.6°.

Exercise 63, page 126: Find the acute angle, to the nearest tenth of a degree, whose cotangent value is 2.127.

Angles whose secant, cosecant, or cotangent values are known can be found using the reciprocals of the cosine, sine, and tangent functions, respectively. Since $\cot\theta = \dfrac{1}{\tan\theta} = 2.127$, we have $\tan\theta = \dfrac{1}{2.127}$, or $(2.127)^{-1}$. To find θ press 1 [ENTER] 2 [·] 1 2 7 [÷] [↰] [ATAN] or 2 [·] 1 2 7 [$1/x$] [↰] [ATAN]. We find that $\theta \approx 25.2°$.

The arccosecant, arcsecant, and arccotangent operations can also be implemented as simple programs:

ACSC: <<INV ASIN>>

ASEC: <<INV ACOS>>

ACOT: <<INV ATAN>>

See Example 7(b) above for more details on saving and running these programs.

CONVERTING BETWEEN DEGREE AND RADIAN MEASURE

We can use the grapher to convert from degree to radian measure and vice versa.

Example 3, page 157: Convert each of the following to radians.

a) 120° b) −297.25°

a) Press [MTH] {{REAL}} [NXT] [NXT]. Then press 1 2 0 [D→R]. The grapher returns a decimal approximation of the radian measure. We see that $120° \approx 2.09$ radians.

b) Press [MTH] {{REAL}} [NXT] [NXT]. Then press 2 9 7 [·] 2 5 [+/−] {{D→R}}. We see that $-297.25° \approx -5.19$ radians.

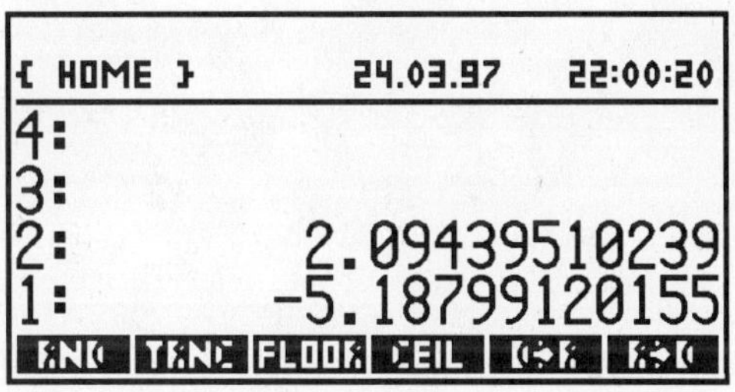

Example 4, page 157: Convert each of the following to degrees.

a) $\frac{3\pi}{4}$ radians b) 8.5 radians

a) Press [MTH] {{REAL}} [NXT] [NXT]. Then press 3 [ENTER] [↰] [π] [×] 4 [÷] {{R→D}} [↰] [→NUM]. (π is the left-shift operation associated with the [SPC] key and →NUM is the left-shift operation associated with the [EVAL] key.) The grapher returns 135, so $\frac{3\pi}{4}$ radians = 135°.

b) Press [MTH] {{REAL}} [NXT] [NXT]. Then press 8 [·] 5 {{R→D}}. The grapher returns 487.014125861, so 8.5 radians ≈ 487.01°.

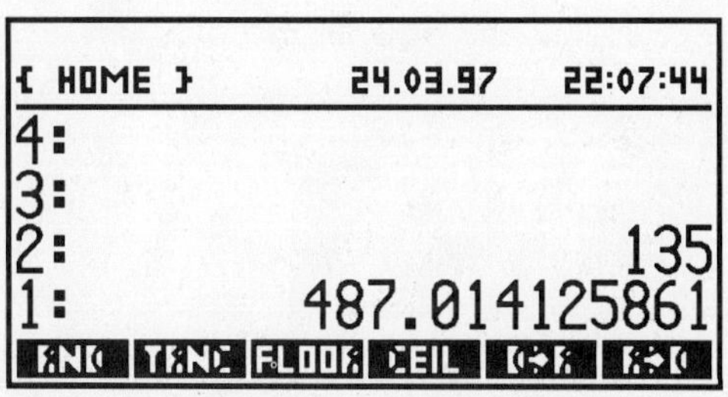

Chapter 3
Trigonometric Identities, Inverse Functions, and Equations

FINDING INVERSE FUNCTION VALUES

We can use a grapher to find inverse function values in both radians and degrees.

Example 2 (a), (e), page 235: Approximate $\cos^{-1}(-0.2689)$ and $\csc^{-1} 8.205$ in both radians and degrees.

To find inverse function values in radians, first set the grapher in Radian mode.

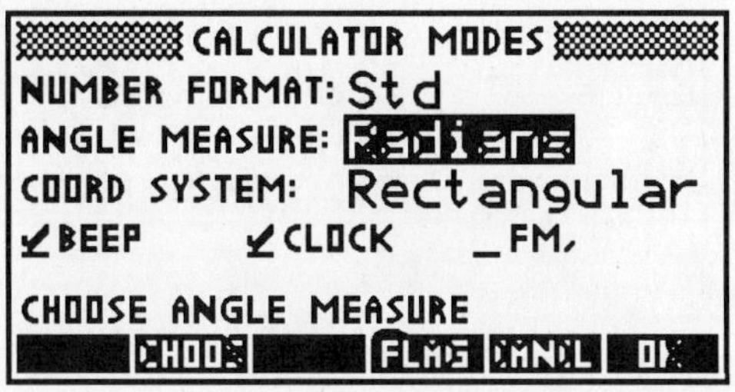

Then to approximate $\cos^{-1}(-0.2689)$, press [·] 2 6 8 9 [+/−] [↰] [ACOS]. The grapher returns 1.84304711148, so $\cos^{-1}(-0.2689) \approx 1.8430$ radians.

To find $\csc^{-1} 8.205$, recall the identity $\csc\theta = \dfrac{1}{\sin\theta}$. Then $\csc^{-1} 8.205 = \sin^{-1}\left(\dfrac{1}{8.205}\right)$. Press 1 [ENTER] 8 [·] 2 0 5 [÷] [↰] [ASIN] or 8 [·] 2 0 5 [$1/x$] [↰] [ASIN]. The readout is .122180665346, so $\csc^{-1} 8.205 \approx 0.1222$ radians.

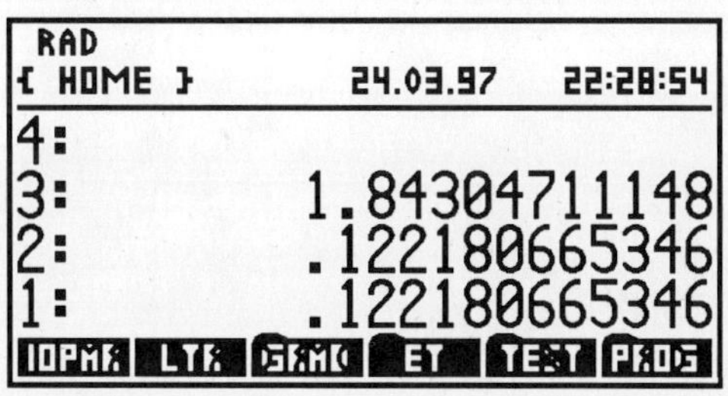

We can also use reciprocal relationships to find function values for arcsecant and arccotangent.

To find inverse function values in degrees, set the grapher in degree mode.

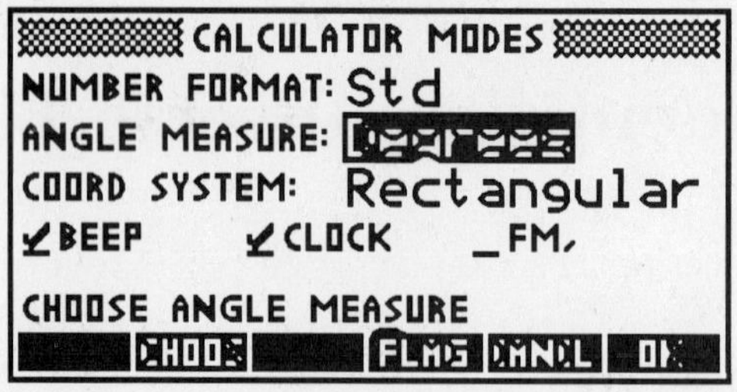

Then use the keystrokes above to find that $\cos^{-1}(-0.2689) \approx 105.6°$ and $\csc^{-1} 8.205 \approx 7.0°$.

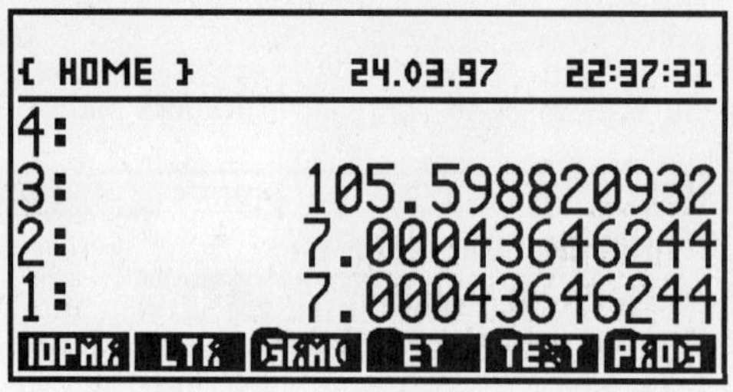

Chapter 4
Applications of Trigonometry

OPERATIONS WITH COMPLEX NUMBERS

Operations with complex numbers can be performed on the HP 48G. On this grapher complex numbers appear as ordered pairs (a, b) where a is the real part and b is the imaginary part. That is, (a, b) represents $a + bi$. A complex number may be entered as an ordered pair, or in the form $a + bi$. The letter i can be obtained by pressing [MTH] and choosing i from the Constant menu, or it can be typed directly with the key sequence [α] [⮢] [CST]. The Equation Writer can also be used to enter expressions with complex numbers.

Example 4, page 281:

(a) Add: $(8 + 6i) + (3 + 2i)$.

To find this sum on the Home screen press [⮢] [( )] 8 [SPC] 6 [ENTER] [⮢] [( )] 3 [SPC] 2 [+]. The result is (11,8), or $11 + 8i$.

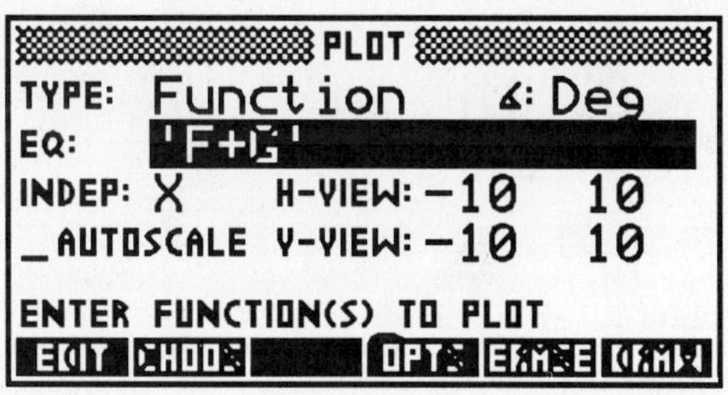

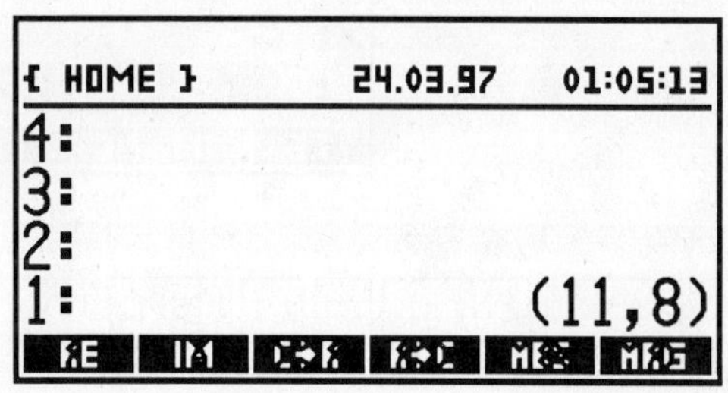

(b) Subtract: $(4 + 5i) - (6 - 3i)$.

We will find this difference using the Equation Writer. Open the Equation Writer. Then press [⮢] [( )] 4 [SPC] 5 [▷] [−] [⮢] [( )] 6 [SPC] [+/−] 3 [ENTER] [EVAL]. The result is $(-2, 8)$, or $-2 + 8i$.

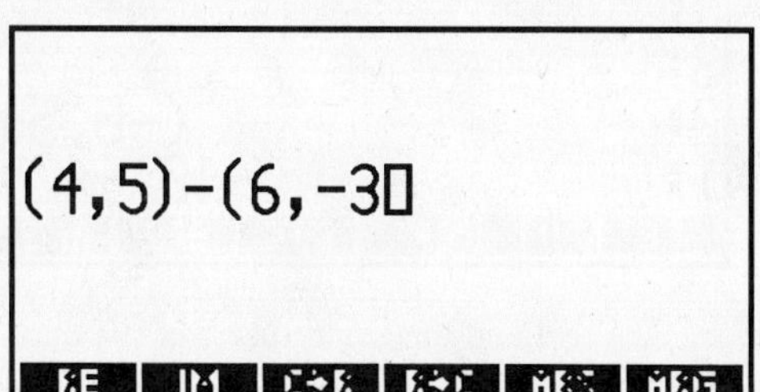

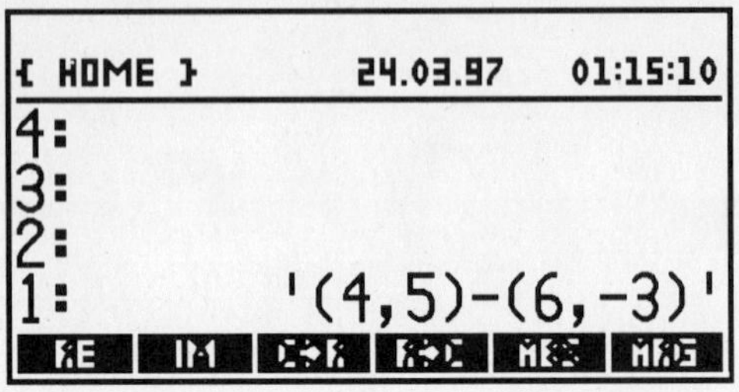

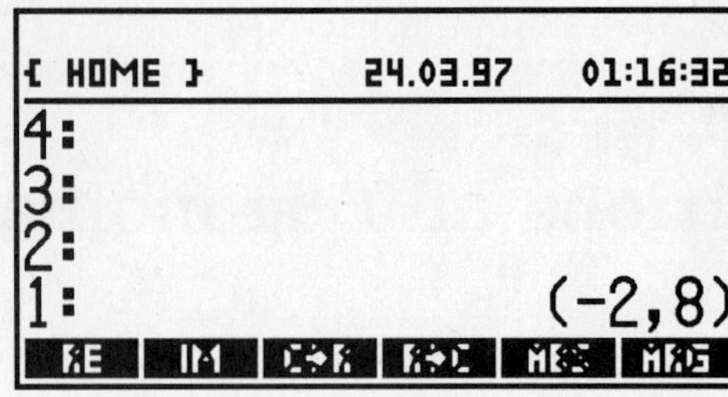

Example 5 (a), (b), (d), (e), page 282:

(a) Multiply: $\sqrt{-16} \cdot \sqrt{-25}$.

On the Home screen press 1 6 [+/−] [$\sqrt{x}$] 2 5 [+/−] [$\sqrt{x}$] [×]. The result is $(-20, 0)$, or -20.

(b) Multiply: $\sqrt{-5} \cdot \sqrt{-7}$.

Press 5 [+/−] [$\sqrt{x}$] 7 [+/−] [$\sqrt{x}$] [×]. Note that this operation produces a decimal approximation of the product. That is, $\sqrt{-5} \cdot \sqrt{-7} \approx -5.91607978309$. The exact value of the product is $-\sqrt{35}$.

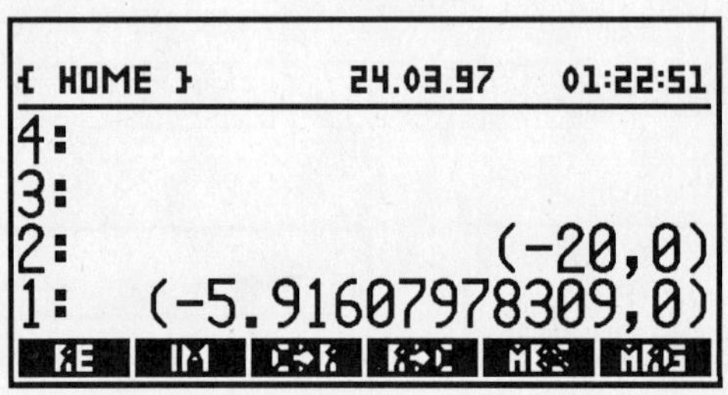

(d) Multiply: $(1 + 2i)(1 + 3i)$.

Press [↰] [( )] 1 [SPC] 2 [ENTER] [↰] [( )] 1 [SPC] 3 [×]. The result is $(-5, 5)$, or $-5 + 5i$.

(e) Multiply: $(3 - 7i)^2$.

Press [↰] [( )] 3 [SPC] 7 [+/−] [↰] [x^2]. The result is $(-40, -42)$, or $-40 - 42i$.

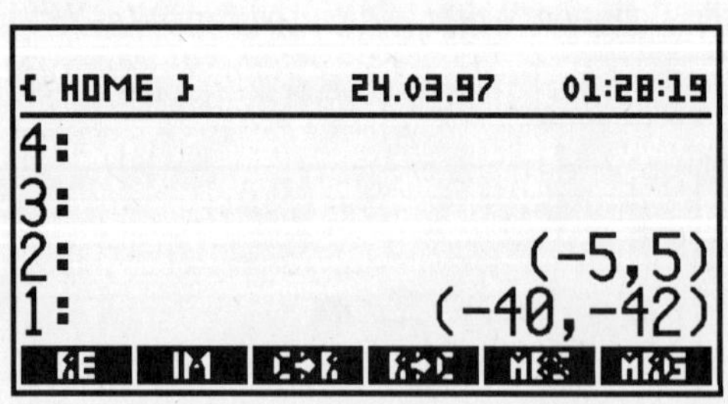

Complex numbers can also be divided. The real and imaginary parts of the quotient will be expressed in decimal form. For example, the quotient $(2 - 5i)/(1 - 6i)$ in Example 9 on page 284 of the text would be expressed as (.864864864865, .189189189189).

Chapter 5
Analytic Geometry Topics

Many conic sections are represented by equations that are not functions. However, the HP 48G can graph these using the Conic Plot type.

GRAPHING PARABOLAS

Example 4, page 328: Graph the parabola $y^2 - 2y - 8x - 31 = 0$.

Open the Plot setup, choose TYPE: Conic, and enter Y∧2 − 2 * Y − 8 * X − 31 = 0 in the EQ: field. There is no need to solve for y. Set the viewing window to $[-12, 12, -8, 8]$ and press {{ERASE}} {{DRAW}},

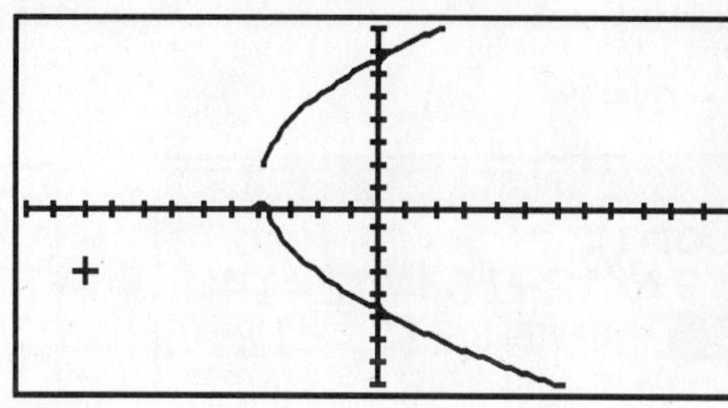

We could also enter the standard equation of the parabola found in the text:

$(y - 1)^2 = 8(x + 4)$.

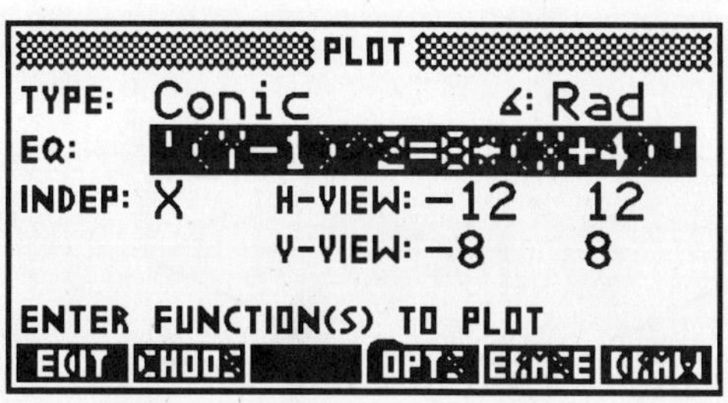

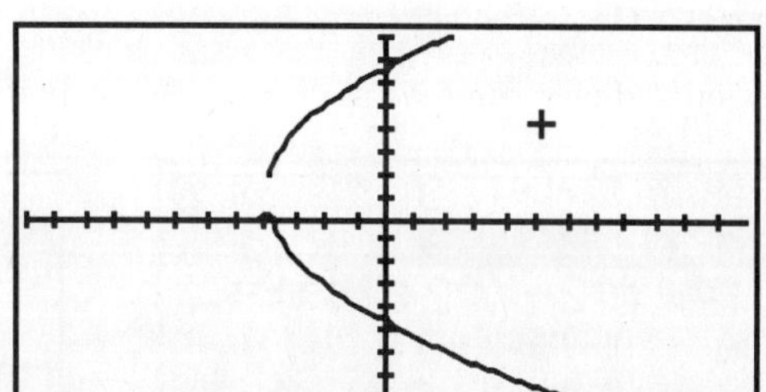

GRAPHING ELLIPSES

In Example 2 of Section 5.2 of the text the procedure for graphing an ellipse of the form $\frac{x^2}{a^2} + \frac{y^2}{b^2} = 1$ or $\frac{x^2}{b^2} + \frac{y^2}{a^2} = 1$ is described. Here we consider ellipses of the form $\frac{(x-h)^2}{a^2} + \frac{(y-k)^2}{b^2} = 1$ or $\frac{(x-h)^2}{b^2} + \frac{(y-k)^2}{a^2} = 1$

Example 4, page 338: Graph the ellipse $4x^2 + y^2 + 24x - 2y + 21 = 0$.

We will enter the equation on the command line of the Plot screen, Highlight EQ: and then press 4 [×] {{X}} [y^x] 2 [+] {{Y}} [y^x] 2 [+] 2 4 [×] {{X}} [−] 2 [×] {{Y}} [+] 2 1 [⬑] [=] 0 [ENTER]. Select a square window (we use $[-12, 12, -8, 8]$) and press {{ERASE}} {{DRAW}}.

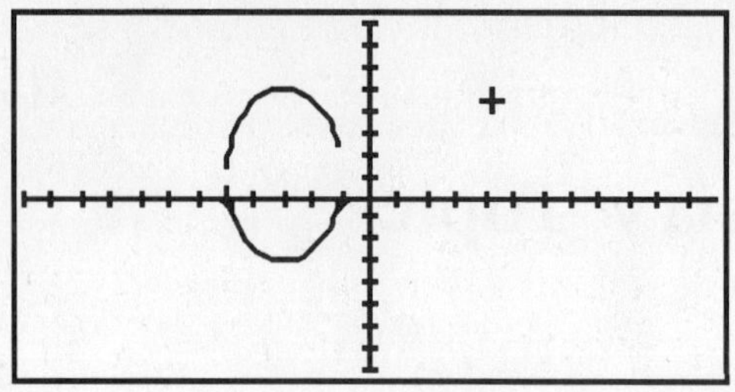

GRAPHING CIRCLES

See Appendix A.2 of this manual (page 163) for a discussion of graphing circles on a grapher.

GRAPHING HYPERBOLAS

As with equations of ellipses, equations of hyperbolas need not be solved for y before they are entered on an HP 48G.

Example 2, page 344: Graph the hyperbola $9x^2 - 16y^2 = 144$.

Enter the equation and plot.

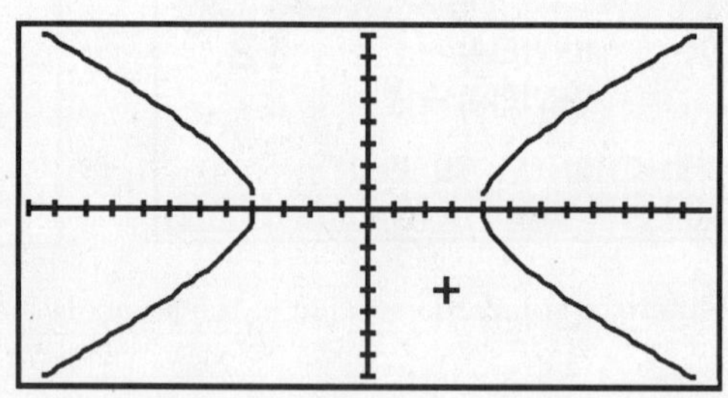

Example 3, page 347: Graph the hyperbola $4y^2 - x^2 + 24y + 4x + 28 = 0$.

Enter the equation and plot. Here we use the window $[-18, 18, -9, 9]$.

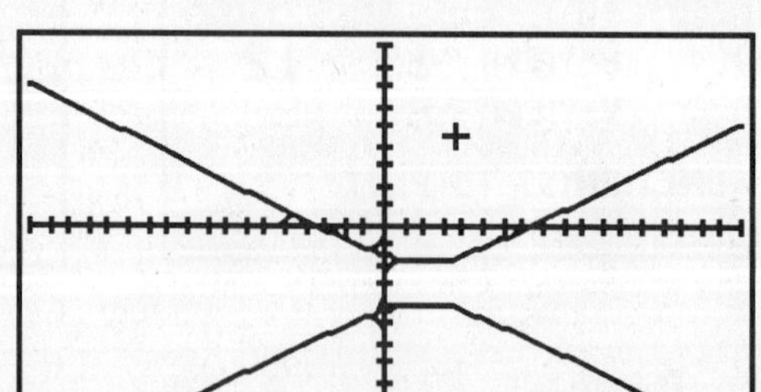

GRAPHING POLAR EQUATIONS

Polar equations can be graphed in radian or degree mode, but the window settings entered must be consistent with the mode. In the Plot setup screen, set TYPE: to Polar. In this example, the grapher is set in radian mode. Typically we begin with a range of $[0, 2\pi]$ or $[0°,360°]$, but it might be necessary to increase the range to ensure that sufficient points are plotted to display the entire graph.

Example 6, page 373: Graph $r = 1 - \sin\theta$.

In the Plot setup, highlight INDEP: and enter θ by pressing [α] [↱] F [ENTER] or by using the CHARS application, [↱] [CHARS]. Highlight EQ: and enter $1 - \sin(\theta)$ by pressing 1 [$-$] [SIN] {{θ}} [ENTER]. Set the window to $[-6, 6 - 3, 3]$. In {{OPTS}}, set LO: to 0 and HI: to 2π. These are, respectively, the lowest and highest values of the independent variable θ to be evaluated. Now, 2π must be in decimal form before it can be entered, so highlight HI:, press [NXT] {{CALC}}, and perform the computation by pressing 2 [ENTER] [↰] [π] [$\times$] [↰] [→NUM]. Then press {{OK}}. The step size is $\pi/24$. This must also be computed and entered in a manner similar to that used for 2π. The default step size is $\pi/90$, or 2°.

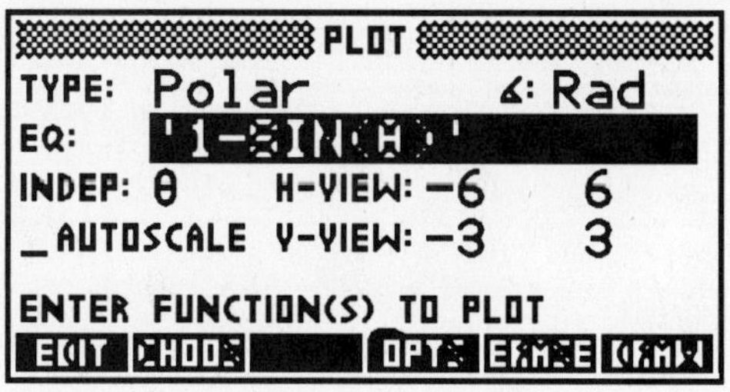

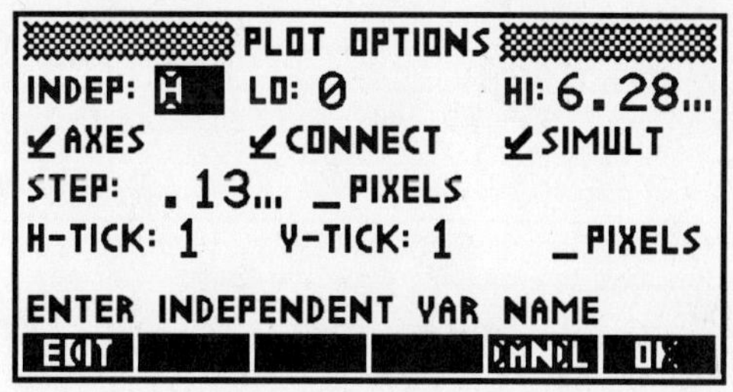

With these settings the grapher evaluates the function from $\theta = 0$ to $\theta = 2\pi$ in increments of $\pi/24$ and displays the graph in the square window $[-6, 6, -3, 3]$. On the Plot screen press {{ERASE}} {{DRAW}} to display the graph.

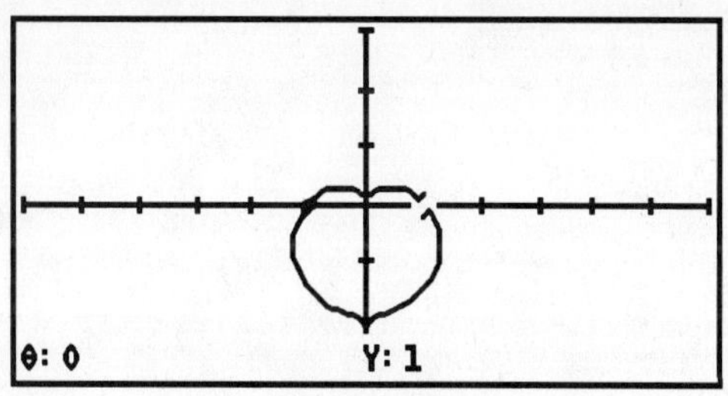

In the figure above, TRACE mode with {{X,Y}} selected is shown. Note that the dependent variable is shown as Y rather that r. The [◁] key will move the cursor around the curve clockwise and [▷] moves it counterclockwise. You can press [▷] indefinitely in TRACE mode, even beyond the maximum value set for θ in Plot options.

Chapter 6
Exponential and Logarithmic Functions

EVALUATING e^x, Log x, and Ln x

Use the grapher's scientific keys to evaluate e^x, $\log x$, and $\ln x$ for specific values of x.

Example 6 (a), (b), page 407: Find the value of e^3 and $e^{-0.23}$. Round to four decimal places.

To find e^3 press 3 [⬑] [e^x]. (e^x is the left-shift operation associated with the [$1/x$] key.) The grapher returns 20.0855369232. Thus, $e^3 \approx 20.0855$. To find $e^{-0.23}$ press [·] 2 3 [+/−] [⬑] [e^x]. The grapher returns .794533602503, so $e^{-0.23} \approx 0.7945$.

Example 4 (a), (b), (c), page 416: Find the values of log 645,778, log 0.0000239, and log (−3). Round to four decimal places.

To find log 645,778 press 6 4 5 7 7 8 [⬏] [LOG]. (LOG is the right-shift operation associated with the [y^x] key.) The grapher returns 5.81008324563. Thus, $\log\ 645,778 \approx 5.8101$. To find log 0.0000239 press [·] 0 0 0 0 2 3 9 [⬏] [LOG]. The grapher returns −4.62160209905, so $\log 0.0000239 \approx -4.6216$. When we press 3 [+/−] [⬏] [LOG] we get a complex number (expressed as an ordered pair). This indicates that −3 is not in the domain of the function $\log x$.

Example 5 (a), (b), (c), page 416: Find the values of ln 645,778, ln 0.0000239, and ln (−5). Round to four decimal places.

To find ln 645,778 and ln 0.0000239 repeat the keystrokes used above to find log 645,778 and log 0.0000239 but press [⬏] [LN] rather than [⬏] [LOG]. (LN is the right-shift operation associated with the [$1/x$] key.) We find that ln 645,778 ≈ 13.3782 and $\ln 0.0000239 \approx -10.6416$. When we press 5 [+/−] [⬏] [LN] we get a complex number,expressed as an ordered pair, indicating that −5 is not in the domain of the function $\ln x$.

USING THE CHANGE OF BASE FORMULA

To find a logarithm with a base other than 10 or e we use the change-of-base formula, $\log_b M = \dfrac{\log_a M}{\log_a b}$, where a and b are any logarithmic bases and M is any positive number.

Example 6, page 417: Find $\log_5 8$ using common logarithms.

We let $a = 10$, $b = 5$, and $M = 8$ and substitute in the change-of-base formula. Press 8 [⬏] [LOG] 5 [⬏] [LOG] [÷]. The result is about 1.2920.

Example 7, page 417: Find $\log_4 31$ using natural logarithms.

We let $a = e$, $b = 4$, and $M = 31$ and substitute in the change-of-base formula. Press 3 1 [⬏] [LN] 4 [⬏] [LN] [÷]. The result is about 2.4771.

Example 9, page 418: Graph $y = \log_5 x$.

To use a grapher we must first change the base to e or 10. Here we use e. Let $a = e$, $b = 5$, and $M = x$ and substitute in the change-of-base formula. Enter $\frac{\ln x}{\ln 5}$ in the Plot setup screen with TYPE: set to Function by pressing [↱] [LN] {{X}} [▷] [÷] [↱] [LN] 5 [ENTER]. Then select a window, and press {{ERASE}} {{DRAW}}.

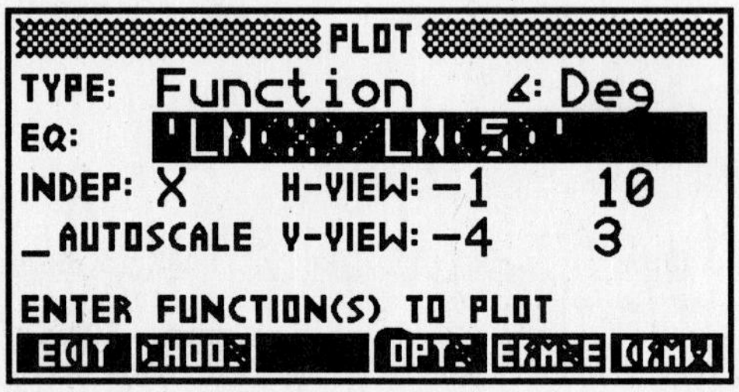

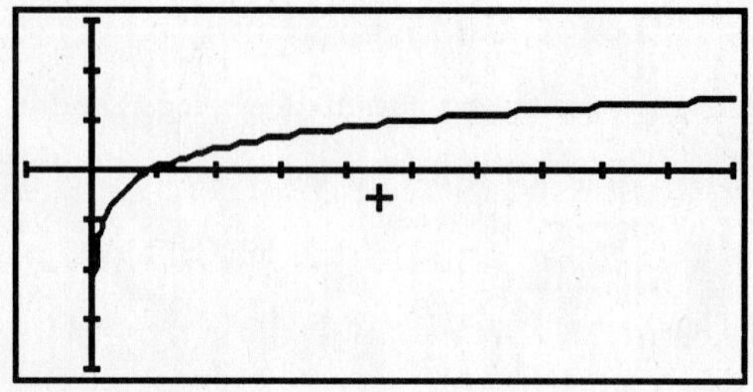

EXPONENTIAL, LOGARITHMIC, AND POWER REGRESSION

Exponential, logarithmic, and power functions can be fit to data. The operations of entering data, making scatterplots, and graphing and evaluating these functions are the same as for linear regression functions. See Appendix A.3 of this manual (page 165) for a discussion of these procedures.

Example 6 (a), page 447: Fit an exponential equation to the given data on cellular phones.

On the Fit Data screen highlight MODEL:, press {{CHOOS}}, and select Exponential Fit. Then enter the data as described on page 165 of this manual. After the last data entry is made, press [ENTER] again followed by {{OK}}. The exponential function appears on the stack on level 3. As before, the entire equation can be viewed by using the arrow keys to move to level 3 and the pressing {{VIEW}}.

Exercise 26 (a), page 453: Fit a logarithmic function to the given data on forgetting.

On the Fit Data screen highlight MODEL:, press {{CHOOS}}, and select Logarithmic Fit. Then enter the data as described on page 165 of this manual. After the last data entry is made, press [ENTER] again followed by {{OK}}. The logarithmic function appears on the stack on level 3. As before, the entire equation can be viewed by using the arrow keys to move to level 3 and the pressing {{VIEW}}.

Example 7 (a), page 449: Fit a power function to the given data on cholesterol level and risk of heart attack.

On the Fit Data screen highlight MODEL:, press {{CHOOS}}, and select Power Fit. Then enter the data as described on page 165 of this manual. After the last data entry is made, press [ENTER] again followed by {{OK}}. The power function appears on the stack on level 3. As before, the entire equation can be viewed by using the arrow keys to move to level 3 and the pressing {{VIEW}}.

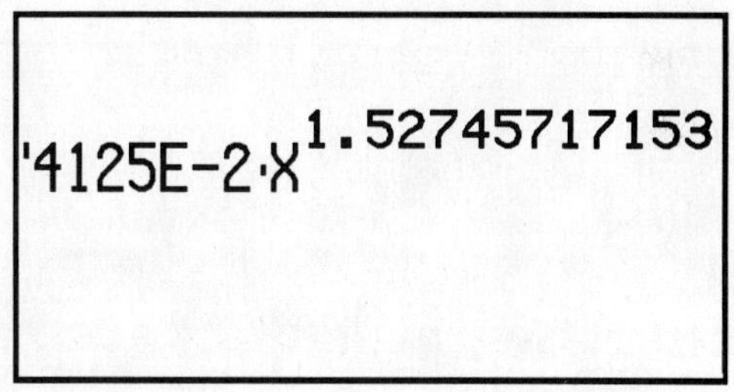

Appendix A.1
Solving Equations

SOLVING QUADRATIC EQUATIONS

The Solve poly. . . feature of the HP 48G's SOLVE operation provides a quick way to solve a quadratic equation written in standard form, $ax^2 + bx + c = 0$.

Example 5, page 464: Solve $3x^2 + 2x = 7$.

First we write the equation in standard form, $3x^2 + 2x - 7 = 0$. Then open Solve by pressing [↱] [SOLVE]. Select Solve poly. . . by pressing [▽] [▽] {{OK}} (or {{ENTER}}). Enter the coefficients of the polynomial, enclosed in brackets, beginning with the coefficient of the term of the highest degree and continuing in descending order. Enter spaces between the coefficients. Remember to enter zeros for missing terms of the polynomial. In this case, press [↰] [[ ]] 3 [SPC] 2 [SPC] 7 [+/−] [ENTER]. Then press {{SOLVE}}. Decimal approximations of the solutions appear in the ROOTS: field. We see that the solutions are approximately 1.230 and −1.897. These are approximations of the exact solutions $\frac{-1+\sqrt{22}}{3}$ and $\frac{-1-\sqrt{22}}{3}$, respectively.

Appendix A.2
Distance, Midpoints, and Circles

GRAPHING CIRCLES

As we saw in Chapter G of this manual, equations of circles can be graphed in the Conic plot type. Square windows can either be entered manually in the Plot setup, or {{ZSQR}} can be pressed in the {{ZOOM}} menu when the graph is plotted. Simply choose Conic in the TYPE: field and enter the equation in the EQ: field. The other fields are similar to those in the Function plot type.

Exercise 37, page 475: Graph $(x-1)^2 + (y-5)^2 = 36$.

We first use the non-square window $[-12, 12, -2, 12]$. The equation does not need to be set equal to zero. To enter the equation highlight the EQ: field and press [↰] [( )] {{X}} [−] 1 [▷] [y^x] 2 [+] [↰] [( )] {{Y}} [−] 5 [▷] [y^x] 2 [↰] [=] 3 6. The press {{ERASE}} {{DRAW}} to see the graph.

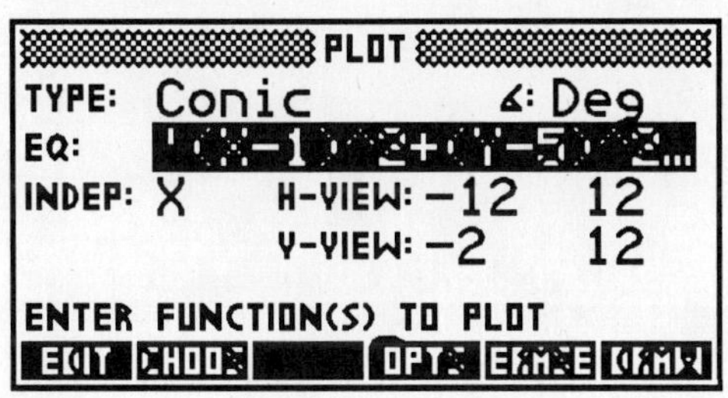

The circle looks distorted.

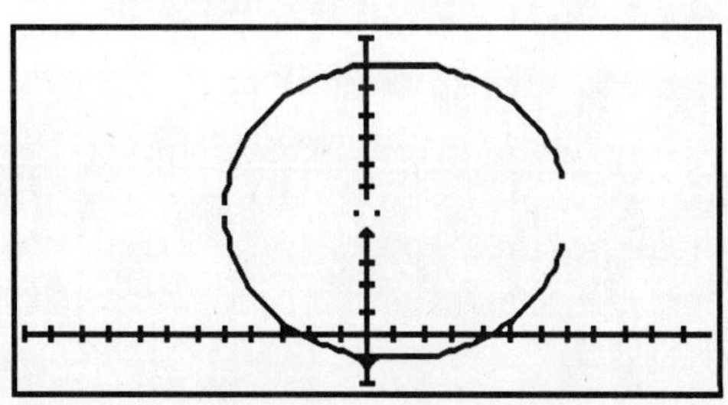

Press {{ZOOM}} {{ZSQR}} to square the window. This produces a graph that is not distorted.

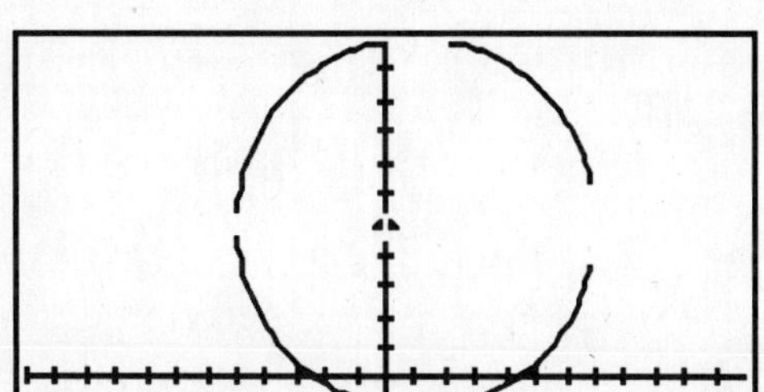

Appendix A.3
Data Analysis, Curve Fitting, and Regression

LINEAR REGRESSION AND SCATTERPLOTS

The linear regression operation of the grapher enables us to fit a linear function to a set of data.

Example 1, page 478: Fit a regression line to the data on life expectancy of women on page 478 of the text. Then predict the life expectancy of women in 2000.

First we enter the data from the table into the Fit Data view of the Statistics application. Press [↱] [STAT] [▽] [▽] {{OK}} to open the Fit Data screen. Enter the independent variable column number (1 in this example) in X-COL: and the dependent variable column number (2 in this example) in Y-COL:. Highlight MODEL:, press {{CHOOS}}, highlight Linear Fit, and press [ENTER].

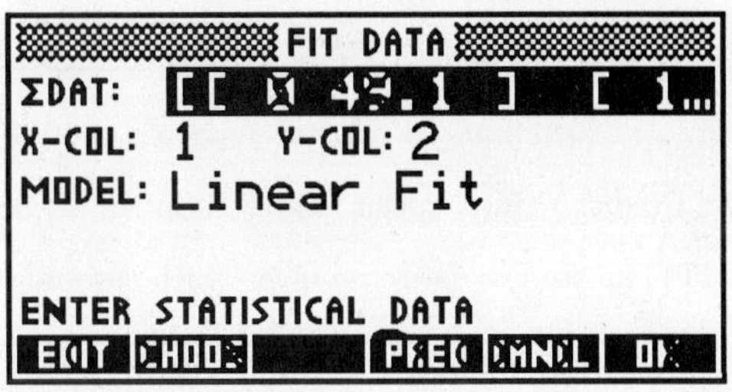

To clear any existing data in the ΣDAT: field press [DEL] {{OK}} after highlighting the field. Press {{EDIT}} to open the screen in which the data will be entered..

As indicated, we will enter the x-values in column 1 and the y-values in column 2. The cursor should be at the top of the left-hand column. Press {{GO↓}} to indicate that the data will be entered going down the columns. Then to enter 0 press 0 [ENTER]. Continue typing the x-values 1 through 9, each followed by [ENTER]. Press [▷] to move to the top of column 2. Type the y-values in succession, each followed by [ENTER].

After all the data is entered press [ENTER] again followed by {{OK}}. The grapher returns the regression model on level 3 of the stack, the correlation coefficient on level 2, and the covariance on level 1. It is likely that the numbers will run off-screen. To view a particular level, use the arrow keys to move to it and press {{VIEW}}.

To view level 3, for example, press [△] [△] [△] {{VIEW}}. The grapher will display the regression equation in the Equation Writer screen. To scroll right or left,use the [▷] or [◁] keys, respectively, and to make a jump to the extreme right or left, press [↱] [▷] or [↱] [◁], respectively. To return to the stack, press [ON] [ON], and press [ON] again to leave the View mode.

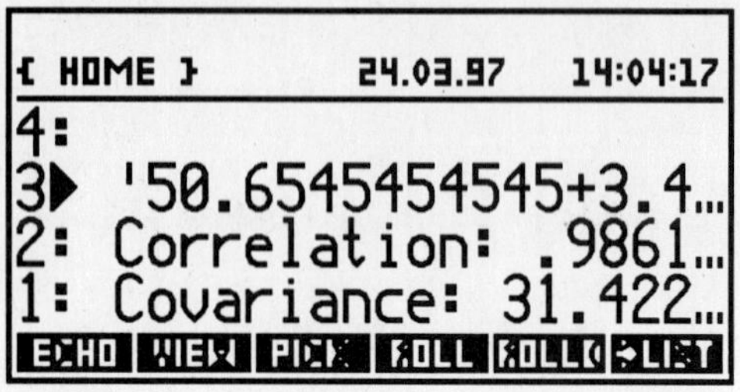

To see a scatterplot of the data, press ⬑ 5 {{PLOT}} {{SCATR}} to plot the data using auto-scaling. With auto-scaling the grapher automatically defines a viewing window that displays all of the data points. This cannot be done before the regression analysis described above. Then press {{STATL}} to see the regression curve along with the scatterplot.

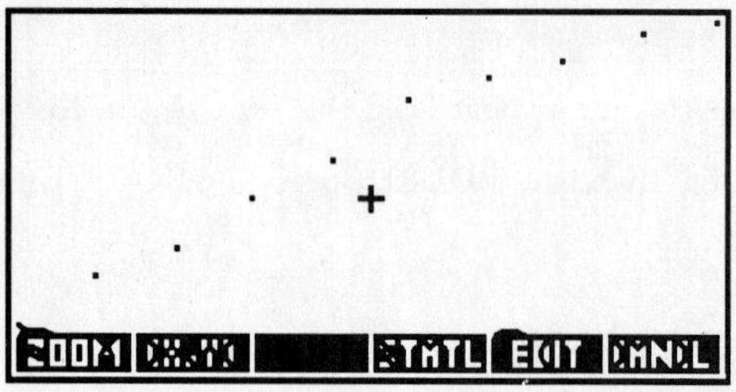

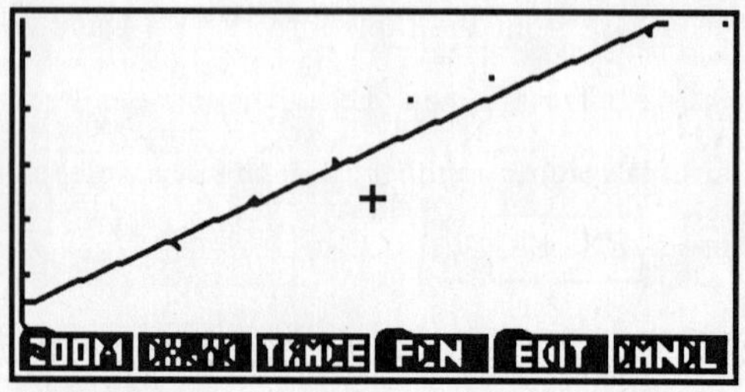

To predict the life expectancy of women in 2000 find the value of y when $x = 10$ in the following manner. In the Fit Data screen, press {{PRED}} to display the Predict Values screen. Enter 10 in the X: field. Press {{PRED}}, highlighting the Y: field first, if necessary. Press {{EDIT}} to see the entire number which results.

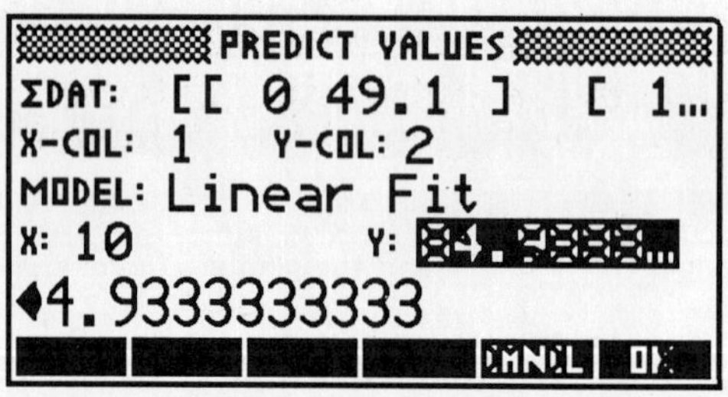

Programs For The HP 48G/48GX Graphic Calculator

PROGRAMS FOR THE HP 48G GRAPHICS CALCULATOR

Created by Michael J. Rosenborg

QUADFORM

It is not necessary to use a program to solve quadratic equations on the HP 48G, because the grapher's Solve poly. . . operation does this. See page 161 of this manual.

TRISOL90

This program solves right triangles *ABC* in which legs *L* and *M* are opposite angles *A* and *B*, respectively. It consists of the main program, TRISOL90, and four subroutines: HA, HL, LA, and LL. It is activated by running TRISOL90 and following the prompts.

INPUT: The hypotenuse and one acute angle (HYP-ANG), the hypotenuse and one leg (HYP-LEG), one leg and one acute angle (LEG-ANG), or both legs (LEG-LEG)

OUTPUT: First the sides; then the acute angles

```
<<''PLEASE PICK ONE''
{ { ''HYP-ANG''
<< HA
>> } { ''HYP-LEG''
<< HL
>> } { ''LEG-ANG''
<< LA
>> } { ''LEG-LEG''
<< LL
>> } } 1 CHOOSE
IF 1 SAME
THEN −17 CF −18
CF EVAL ''L1 = '' L
→STR + ''L2 = '' M
→STR + ''HYP = '' H
→STR + ''ANG 1 = '' A
→STR + ''ANG 2 = '' B
→STR + 5 →LIST { L
M H A B } PURGE
END
>>
```

HA:

```
<< ''HYP-ANG'' { {
''HYPOTENUSE:''
''ENTER LENGTH'' 0 }
{ ''ANGLE:''
''ENTER DEGREES'' 0 }
} { } { } { }
INFORM
IF 1 SAME
THEN OBJ→ DROP
'A' STO 'H' STO 90
A − 'B' STO H A SIN
* 'L' STO H B SIN *
'M' STO
END
>>
```

HL:

```
<< ''HYP-LEG'' { {
''HYPOTENUSE:''
''ENTER LENGTH'' 0 }
{ ''LEG:''
''ENTER LENGTH'' 0 }
} { } { } { }
INFORM
IF 1 SAME
THEN OBJ→ DROP
'L' STO 'H' STO H
SQ L SQ - √ 'M' STO
L H / ASIN 'A' STO
90 A - 'B' STO
END
>>
```

LA:

```
<< ''LEG-ANG'' { {
''LEG:''
''ENTER LENGTH'' 0 }
{ ''ANGLE:''
''ENTER DEGREES'' 0 }
} { } { } { }
INFORM
IF 1 SAME
THEN OBJ→ DROP
'A' STO 'L' STO 90
A - 'B' STO L A TAN
/ 'M' STO L SQ M SQ
+ √ 'H' STO
END
>>
```

LL:

```
<< ''LEG-LEG'' { {
''LEG 1:''
''ENTER LENGTH'' 0 }
{ ''LEG 2:''
''ENTER LENGTH'' 0 }
} { } { } { }
INFORM
IF 1 SAME
THEN OBJ→ DROP
'M' STO 'L' STO L
SQ M SQ + √ 'H' STO
L M / ATAN 'A' STO
90 A - 'B' STO
END
>>
```

TRISOL

This program solves triangle ABC, in which sides L, M, and N are opposite angles A, B, and C, respectively. It covers the "no solution" case as well as the "two solutions" case. The program consists of the main program, TRISOL, and six subroutines: INIT, SSS, SAS, AAS, ASA, and SSA. It is activated by running TRISOL and following the prompts.

INPUT: Three parts of a triangle, including at least one side
OUTPUT: First the sides, then the angles

TRISOL:

```
<< INIT
IF 1 FS?C
THEN      IF B A ≥
THEN { A B C L
M N } PURGE KILL
ELSE
''2 SOLUTIONS . . .''
MSGBOX 180 A – 'A'
STO 180 A – B – 'C'
STO C SIN M * B SIN
/ 'N' STO
''SIDE a = '' L →STR
+ ''SIDE b = '' M
→STR + ''SIDE c = ''
N →STR +
''ANGLE A = '' A →STR
+ ''ANGLE B = '' B
→STR + ''ANGLE C = ''
C →STR + 6 →LIST
END { A B C L M
N } PURGE
END { A B C L M N
} PURGE 1 CF
>>
```

INIT:

```
<< 1 CF
''PLEASE PICK ONE'' {
{ ''SSS''
<< SSS
>> } { ''SAS''
<< SAS
>> } { ''AAS''
<< AAS
>> } { ''ASA''
<< ASA
>> } { ''SSA''
<< 1 SF SSA
>> } } 1 CHOOSE
IF 1 SAME
THEN –17 CF –18
CF EVAL
END ''SIDE a = '' L
→STR + ''SIDE b = ''
M →STR +
```

```
''SIDE c = '' N →STR
+ ''ANGLE A = '' A
→STR + ''ANGLE B = ''
B →STR +
''ANGLE C = '' C →STR
+ 6 →LIST
>>
```

SSS:

```
<< ''SSS'' { {
''SIDE a:''
''ENTER LENGTH'' 0 }
{ ''SIDE b:''
''ENTER LENGTH'' 0 }
{ ''SIDE c:''
''ENTER LENGTH'' 0 }
} { } { } { }
INFORM
IF 1 SAME
THEN OBJ→ DROP
'N' STO 'M' STO 'L'
STO M SQ N SQ + L
SQ - 2 M * N * /
ACOS 'A' STO L SQ N
SQ + M SQ - 2 L * N
* / ACOS 'B' STO
180 A - B - 'C' STO
END
>>
```

SAS:

```
<< ''SAS'' { {
''SIDE a:''
''ENTER LENGTH'' 0 }
{ ''ANGLE C:''
''ENTER DEGREES'' 0 }
{ ''SIDE b:''
''ENTER LENGTH'' 0 }
} { } { } { }
INFORM
IF 1 SAME
THEN OBJ→ DROP
'M' STO 'C' STO 'L'
STO L SQ M SQ + 2 L
* M * C COS * - √
'N' STO M SQ N SQ +
L SQ - 2 M * N * /
ACOS 'A' STO 180 A
- C - 'B' STO
END
>>
```

AAS:

```
<< ''AAS'' { {
''ANGLE A:''
''ENTER DEGREES'' 0 }
{ ''ANGLE B:''
''ENTER DEGREES'' 0 }
{ ''SIDE a:''
''ENTER LENGTH'' 0 }
} { } { } { }
INFORM
IF 1 SAME
THEN OBJ→ DROP
'L' STO 'B' STO 'A'
STO 180 A - B - 'C'
STO L B SIN * A SIN
/ 'M' STO L C SIN *
A SIN / 'N' STO
END
>>
```

ASA:

```
<< ''ASA'' { {
''Angle A:''
''ENTER DEGREES'' 0 }
{ ''SIDE c:''
''ENTER LENGTH'' 0 }
{ ''ANGLE B:''
''ENTER DEGREES'' 0 }
} { } { } { }
INFORM
IF 1 SAME
THEN OBJ→ DROP
'B' STO 'N' STO 'A'
STO 180 A - B - 'C'
STO N A SIN * C SIN
/ 'L' STO N B SIN *
C SIN / 'M' STO
END
>>
```

SSA:

```
<< ''SSA'' { {
''SIDE a:''
''ENTER LENGTH'' 0 }
{ ''SIDE b:''
''ENTER LENGTH'' 0 }
{ ''ANGLE B:''
''ENTER DEGREES'' 0 }
} { } { } { }
INFORM
IF 1 SAME
THEN OBJ→ DROP
'B' STO 'M' STO 'L'
STO L B SIN * M / →
X

      <<
IF X 1 >
THEN
''NO SOLN.'' MSGBOX {
L M B } PURGE 1 CF
KILL
ELSE X ASIN
'A' STO 180 A - B -
'C' STO M C SIN * B
SIN / 'N' STO
END
>>
END
>>
```

DEMOIVRE

This program helps visualize Nth roots of a complex number $A + Bi$.

INPUT: A complex number $A + Bi$, entered as (A, B), and the index N
OUPUT: Traceable graph of the N points (A, B) corresponding to the roots

```
<<
''DEMOIVRE'S THEOREM''
{ { ''COMPLEX NO.:''
''PRESS NXT {{CALC}} IF NECESSARY''
1 } { ''ROOT INDEX:''
''PRESS NXT {{CALC}} IF NECESSARY''
0 } } { } { } { }
INFORM
IF 1 SAME
THEN OBJ→ DROP
'N' STO C→R 'B' STO
'A' STO A SQ B SQ +
√ N INV ∧ 'R' STO
−16 CF −17 CF −18
CF
CASE A 0 SAME
THEN 180 B
ABS B / 90 * −
END A 0 <
THEN B A /
ATAN 180 +
END B A /
ATAN
END N / 'H' STO
'(R*COS(T),R*SIN(T)
)' 'EQ' STO R 1.5 *
'YMAX' STO YMAX −1
* 'YMIN' STO YMAX
1.5 * 'XMAX' STO
XMAX −1 * 'XMIN'
STO XMIN YMIN R→C
XMAX YMAX R→C T H H
360 + 3 →LIST 360 N
/ { (0,0) { 1 1 } }
''PARAMETRIC'' 'R' 7
→LIST 'PPAR' STO
PARAMETRIC ERASE
DRAW DRAX PICTURE
END { N B A R H
YMAX YMIN XMAX XMIN
} PURGE
>>
```

LOGBASE (PLAIN)

This program computes the logarithm of a positive number to any acceptable base. Enter the number and then the base on the stack before running the program. The logarithm is returned on level 1 of the stack.

```
<<→ X B 'LN(X)/LN(B)' >>
```

LOGBASE (FANCY)

This program computes the logarithm of a positive number X to any acceptable base.
INPUT: X, BASE
OUTPUT: LOGARITHM of X to the given base

```
<< ''LOG BASE CHANGE''
{ { ''X:''
''ENTER NO. > 0'' 0 }
{ ''BASE:''
''ENTER BASE (POS.; ≠ 1)''
0 } } { } { } { }
INFORM
IF 1 SAME
THEN OBJ→ DROP LN
INV SWAP LN *
END
>>
```